AI POWERED UX RESEARCH

Run Research at the *Speed* Your Team Actually Needs

A PRACTITIONER'S FRAMEWORK FOR BUILDING THE RESEARCH ENGINE THAT CLOSES THE GAP BETWEEN DECISION VELOCITY AND EVIDENCE VELOCITY.

CONSTANTINE PAPAS

THEVOICEOFUSER.COM

Published by The Voice of User

ISBN Paperback: 979-8-9967856-0-5

ISBN EPUB: 979-8-9967856-1-2

Contents

CHAPTER 4

CHAPTER 5

CHAPTER 6

CHAPTER 7

CHAPTER 8

CHAPTER 9

CHAPTER 10

CHAPTER 11

A Note from the Author

I did not set out to write a book. I set out to write a blog post.

I run a blog called The Voice of User (thevoiceofuser.com). I started after I was laid off in early 2025, and it grew into something I did not expect: a place where I could say what I actually believe about UX research, rather than what looks good on a conference slide. No corporate framing. No consultant-speak. Just what I've seen after fourteen years of doing this work at companies—and from the other side of the table, as an academic studying how people interact with technology.

In late 2025, I started writing about AI in research—not the conference keynote version, but the one people talk about honestly afterward. The practitioner's version: what these tools actually do, where they break, and what changes when you take them seriously instead of either worshipping them or pretending they will go away.

The posts kept getting longer. The series kept growing. At some point, I realized I was not writing blog posts anymore. I was building a system. That is not a blog series. That is a book.

So I wrote the book.

What this book introduces is a framework I call the Research Engine. It's an operating model built around three research modes: a routing logic for matching questions to the right mode, a concept called the Frame that determines whether fast research is useful or just fast, and a governance structure that keeps the whole system from collapsing under organizational pressure. It is not a set of tool recommendations. It is not a maturity model. It is a practical system for structuring and running a research function in a world where AI has changed what is possible. At its core, it argues that research functions need to move from a service model to an intelligence function, and it lays out the operational system that makes that transition possible.

You do not have to implement everything at once. The system described here is modular by design. A solo researcher at an early-stage startup will begin differently than a research director at a large technology company. Some readers will need the routing logic immediately and the governance infrastructure later. Others will start with the Frame and build outward. The book is written sequentially because the concepts build on one another, but it is designed to be used selectively, based on where your organization actually is.

What kept pushing me forward was a conviction I could not shake: new tooling without new organizational thinking is just old problems running faster. The field was adopting AI tools at speed while leaving the operating model, the governance, and the way research functions are structured and run completely untouched. The tools changed; everything around them stayed the same. That gap is where most of the dysfunction lives, and it is the gap this book aims to close.

I am not neutral about any of this. I have opinions. Fourteen years of being in the room will do that to you. And I could not just watch what was happening with AI adoption.

The pattern was visible from a mile away—the same thing that happened with design thinking, the same thing that happened with agile, the same thing that has happened with every wave of "transformation" that has rolled through the industry over the past two decades. Consultants brief VPs. Frameworks get named and turned into slide decks. Keynotes happen. LinkedIn fills up with thought leaders who have never actually done the work, explaining to people who are doing the work what the work should look like. Then organizations slap the new thing on top of the old structure, call it innovation, measure nothing, and wonder why the productivity gains never materialize.

That is exactly what is happening with AI in tech right now.

I wrote this book because the field deserves better than what it is getting from most of the people who have appointed themselves to explain AI—and because I got tired of waiting for someone else to say so.

I wrote this book for the same people I write my blog for: the researchers who kept doing the work while everyone else debated whether the work still mattered; the ones who sat through meetings where someone said, "We don't have time for research," and then watched the team spend three months building something nobody wanted; the ones who were asked to "just validate this real quick" and had the discipline to say, "That is not what this study can tell you"; and the ones who watched AI arrive and did not panic or dismiss it, but picked it up, figured out what it could actually do, and got back to work.

This book is for those researchers—the ones who are already leading, whether anyone has given them the title yet or not.

— *Constantine Papas*

CHAPTER 1

How AI Is Changing Tech

While preparing this book, I interviewed more than fifty practitioners across product, design, engineering, and research roles at companies ranging from early-stage startups to large, publicly traded technology organizations. I conducted multiple rounds of follow-up conversations to pressure-test what I was hearing. One theme emerged in nearly every conversation, regardless of company size, industry, or team structure: the speed at which product decisions are being made has changed fundamentally, and research has not kept pace.

The consistency of this finding was striking. I expected to hear different stories from different kinds of organizations. I expected startups to talk about speed and enterprises to talk about process. Instead, I heard the same structural tension described from different seats at the table. Researchers talked about being outpaced. PMs talked about making decisions without evidence and feeling uneasy about it. Designers talked about generating more options than anyone could meaningfully evaluate. Engineering leads talked about shipping things faster than anyone could verify whether they were the right things to ship. Everyone was describing the same elephant from a different angle.

This chapter is about that change—not the hype version of it, not the conference keynote version where AI transforms everything and the future arrives on schedule, but the version people described when they were being honest about what their day actually looks like.

The Loop Got Faster. Research Did Not.

The product development loop has been compressing for years. Agile, continuous delivery, feature flagging, and the broader industry realization that shipping small and learning fast beats planning large and hoping hard that everyone contributed. But AI accelerated that compression in ways that are qualitatively different from what came before. It did so by attacking the parts of the loop that used to create natural breathing room for evaluation.

Design used to be a bottleneck—not because designers were slow but because producing high-fidelity mocks and interactive prototypes took real time. A designer could explore maybe three to five directions in a sprint. Now, with generative design tools and AI-assisted prototyping, teams describe designers generating dozens of options in a couple of days. One senior designer at a midsize e-commerce company told me the exploration phase that used to take a sprint now takes a day, sometimes less. The constraint shifted from "How many options can we explore?" to "Which of these options do we actually evaluate?" That is a fundamentally different constraint, and most teams have not adjusted their processes to reflect it.

Engineering followed the same pattern. AI-assisted coding tools do not replace engineers, but they compress the gap between "we know what to build" and "it is built enough to test": scaffolding, boilerplate, test generation, documentation. Multiple engineering leads I spoke with described prototype timelines shrinking to days. The practical effect is that the time between a decision being made and something being in production has compressed significantly. When build cycles are shorter, the cost of slow decisions goes up. A team that can ship in days feels the weight of a two-week research cycle differently than a team that ships quarterly. The mismatch is not dramatic; it is just persistent enough that teams start routing around research instead of waiting for it.

Product managers, meanwhile, are working with AI-generated artifacts at a volume that did not exist two years ago: strategy drafts, competitive analyses, user journey hypotheses, PRD outlines. The quality varies, but the volume is real. PMs can now produce the appearance of having thought deeply about something in a fraction of the time it used to take. Whether they actually thought deeply is a different question—and one I will return to. The point is that the artifacts that used to gate decisions (the spec, the mock, the competitive analysis) are now produced so quickly that they no longer serve as natural speed bumps in the process. The gates that used to slow things down enough for research to contribute have opened.

A product lead at a midstage fintech company told me something that stuck with me. She said her team used to have a natural three- to four-week window between idea and commitment where research could contribute. Now the window is one to two weeks. "The gap did not disappear," she said. "It just got so small that the old research model cannot fit inside it anymore. So we stopped trying, and then we stopped asking." That sentiment should worry every research leader. Not "we do not value research," but "we stopped asking because we assumed you could not help quickly enough."

Nearly every researcher I interviewed described the same experience from different angles. The loop they were designed to fit inside got faster, and nobody rebuilt the research function to match—not because researchers are oblivious, but because the infrastructure, systems, operating models, and organizational expectations around research were built for the old tempo.

What "Fast" Actually Means Now

"Fast" is one of those words people use without defining, and then end up arguing about different things while nodding at each other. In my interviews, I heard a consistent miscommunication that plays out across organizations. A researcher says, "We turned this around fast," meaning two weeks. A PM says, "That was not fast enough," meaning they needed it in two days. Neither realizes they are using entirely different definitions of the same word. That gap is both awkward and structurally dangerous because it leads to broken expectations, missed decision windows, and, eventually, the conclusion that research is too slow to be useful.

Two years ago, a fast product cycle meant shipping a meaningful feature update every five to seven weeks. A fast research cycle meant delivering a study in two to three weeks instead of six. These timelines were considered aggressive.

Today, "fast" means something different. A fast product cycle means iterating on a feature daily or near daily, using feature flags, targeted rollouts, and rapid experiments to learn in production. A fast decision cycle means resolving a product question in hours or days, not weeks. Teams are not just shipping faster; they are deciding faster. The unit of work has shrunk. Instead of large, multiweek feature bets, teams are making dozens of small, scoped decisions per sprint: this copy or that copy, this flow or that flow, this default or that default, surface this information or hide it.

A head of product at a B2B SaaS company described the shift in terms that were hard to argue with. Two years ago, her team made maybe five significant product decisions per quarter that could benefit from user evidence. Now, with faster prototyping and shorter build cycles, that number is closer to fifteen or twenty. The decisions got smaller, but they also got more frequent, and the research team did not triple in size to match.

Each of those decisions is small on its own. But they compound. And each one is a moment when the team either has evidence or does not—either has some signal about what users actually experience or guesses based on internal opinion and pattern matching. The total number of decision points in a single product quarter has gone up significantly, while the time available to inform any single decision has gone down.

This is the part that most discussions about AI and product development skip over. The story is not just that teams ship faster. The story is that the decision surface area has expanded while the decision window has contracted: more choices to make, less time to make each one, and the same number of researchers (or fewer, after several rounds of layoffs while the industry was busy talking about AI efficiency).

A researcher at a large consumer technology company described it this way during one of my interviews. A PM had a question about whether users understood the difference between two similar features displayed on the same screen. In the old model, this would have been queued as a study, scoped over a few days, recruited over a week, run over another week, and synthesized for a readout two to three weeks later—by which point the team would have shipped something anyway, because the sprint cadence does not wait for research. It never did; everyone just pretended it did.

That researcher had built systems that allowed them to turn the question around in two days, with real participants and explicit boundaries on what the findings could and could not support. The team was surprised this was possible, and that surprise is telling. Their mental model of research was still calibrated to the old tempo. They had already prepared to pick an option based on internal debate and move on. Research was not even on the table—not because they did not value it but because they assumed it could not operate at the speed they needed.

And here is the part that stings. The teams that move fastest are often not ignoring research. They are the ones that have figured out how to get evidence at the speed they need. Some have done it well, with real methods, real participants, and honest constraints on their conclusions. But many have done it badly: with quick surveys that ask leading questions, with AI-generated synthesis that nobody audits, with "research" that is actually a product manager asking a friend whether a flow makes sense. The absence of a credible fast-research operating model does not mean teams stop seeking evidence. It means they seek bad evidence and treat it as good enough—which is, in some ways, worse than no evidence at all, because at least with nothing you know you are guessing.

Decision Volume Is Exploding. Decision Quality Is Not Keeping Up.

Here is the math. The number of product decisions that could benefit from user evidence has increased by an order of magnitude. The number of researchers available to generate that evidence has stayed flat or decreased. The time available per decision has shrunk. And the tools AI gave product teams to move faster did not come with corresponding tools for researchers to evaluate faster—at least not automatically, and not without serious judgment about when and how to use them.

The result is a gap—a large, growing, structurally persistent gap between decision velocity and evidence velocity. That gap gets filled with something. It always does. A product team with a deadline and a VP asking for progress updates will not sit in silence waiting for data.

What fills the gap? I heard the same patterns across nearly every organization I spoke with.

First, internal opinion dressed up as user insight. Someone senior says, "I think users would prefer this," and because there is no time to check, that opinion becomes the operating assumption. It gets written into the spec, shapes the design, and becomes load-bearing truth—and it was never tested. Nobody says, "We are guessing." They say, "Based on our understanding of user behavior," or "Consistent with what we have seen," or "Users want simplicity." Everyone agrees, because who is going to argue with simplicity? Nobody asked a user. The feature ships. If it works, the opinion gets reinforced as wisdom. If it fails, the economics get blamed, or the market, or "adoption headwinds." The opinion itself is protected. Several PMs I spoke with were refreshingly honest about this pattern. One said, "We call it intuition, but it is really just the loudest person in the room."

Second, analytics misread as user understanding. Teams look at funnel data, click rates, and engagement metrics and conclude that they understand why users behave the way they do. They do not. They understand what users did. They have no idea why. A 40 percent drop-off at step 3 of a flow could mean the step is confusing, or it could mean users accomplished their goal at step 2 and did not need step 3 at all. Those are opposite interpretations of the same number, and both are plausible. Without qualitative signal, the interpretation is a coin flip. But the chart looks authoritative in a meeting, nobody has time to question it, and the analyst presenting it has no incentive to add the caveat "we actually have no idea why."

A data scientist at an enterprise SaaS company described this to me as the most common failure mode on his team. "We can tell you what happened with high precision," he said, "but teams treat the what as if it explains the why. They see a drop and immediately start building solutions before anyone has talked to a single user about what they experienced. Half the time, the problem they are solving is not the problem users actually have." He described a case where the team spent a full quarter optimizing a flow based on funnel data, only to discover through later user interviews that the real issue was terminology confusion on the first screen. The data showed where people left; it did not show that they were confused about what they were looking at in the first place.

Third, AI-generated insight theater. I wrote about this extensively on my blog. This is the newest and, in some ways, most dangerous gap filler. Teams talk to a few users, paste transcripts into an LLM, ask it to "find the themes," get five confident-sounding bullet points, put them in a deck, and present them as findings. There is no evidence trail, no link between a claim and a specific participant moment, no audit, no check for disconfirming evidence. If someone asks, "Where did this insight come from?", the honest answer is, "The AI said so, and it sounded right." That is not research. Multiple researchers I interviewed described finding this exact pattern in their organizations and having no clear mechanism to push back, because the output looked polished and the speed was exactly what leadership wanted.

Fourth—and this is the silent killer—decision debt. When there is no evidence and no one wants to guess out loud, decisions simply do not get made. They get deferred, pushed to the next sprint, or tabled for "when we have more data." The feature ships with a default that nobody chose deliberately, and that default becomes permanent because changing it requires a new decision—and there is no evidence for that one either. Decision debt compounds like financial debt: quietly, invisibly, until it becomes structural. Six months later, someone asks why the product feels incoherent, and nobody has a good answer, because the incoherence was not the result of one bad decision. It was the result of dozens of nondecisions accumulating into a product nobody designed on purpose.

I asked multiple teams in my interviews to trace a recent product problem back to its origin. In almost every case, the origin was not a bad decision. It was a nondecision: a default that stuck, a question nobody asked because asking it would have meant slowing down. A PM at a consumer health tech company described a feature that had been accumulating complaints for months. When the team finally investigated, it discovered the root cause was a content hierarchy decision made eight months earlier. Nobody had actually chosen the hierarchy. It was a placeholder from an early prototype that shipped because changing it would have required user input nobody had time to gather. Eight months of user confusion because a two-day study never happened.

The pattern is consistent. When evidence cannot keep up with decision velocity, quality degrades—not dramatically, not in ways that produce immediate, visible failures, but in slow, accumulating ways: small wrong bets, slight misreads of user behavior, features that are 80 percent right but 20 percent misaligned in ways nobody notices until retention softens six months later and no one can trace back to any single cause.

This is the environment this book is written for. Not a crisis, but something quieter: a slow, normalized erosion of decision quality that people across my interviews could feel but could not always point to directly.

And to be clear, this is not a moral failing by anyone involved. Nobody in this scenario is being irresponsible. The PM who relies on opinion is doing so because the alternative is missing a launch window. The analyst who overinterprets funnel data is doing so because the team needs a narrative and the data is what it has. The researcher who lets AI summarize their transcripts is doing so because they have eleven studies in flight and four hands. Everyone is responding rationally to a system that creates more decisions than it can support with evidence. The problem is structural, not personal—which means the solution has to be structural too.

Generating Options Faster Than You Can Evaluate Them

One of the most telling stories from my interviews came from a design lead at a consumer fintech company. Their team had started using an AI prototyping tool, and a designer generated twelve distinct layout variations for a feature screen in about twenty minutes—not rough sketches, but twelve plausible options. A year earlier, producing twelve options of that quality would have taken a full sprint.

The team looked at the twelve options. The PM said, "These all look good." The designer had favorites. The engineer flagged which ones were easiest to build. Nobody suggested checking which ones users could actually navigate—not because they did not care about users, but because there was no process for getting a real answer to that question quickly enough before committing to a direction. The infrastructure for generating options had outpaced the infrastructure for evaluating them.

They shipped one of the options. It was fine—not wrong, not broken, just fine. That word should concern you more than "failed," because fine means nobody will ever go back and ask whether it could have been better. Fine is the sound of a team that stopped looking for the best answer because a good-enough answer was available and time was short. Multiply that by every decision, every sprint, every team in an organization, and the compounding cost becomes real.

I heard versions of this story repeatedly. The rate of option generation has increased across every function. AI allows PMs to draft more specs, designers to explore more directions, engineers to scaffold more prototypes, and marketers to produce more copy variants. The generative parts of product development have been supercharged. The evaluative parts have not.

To put it plainly: AI made it cheap to propose. It did not make it any cheaper to validate. Every option generated without evaluation is a small bet placed without evidence. The more options generated, the more bets placed—and the more important it becomes to have a fast, credible way to check at least some of them against reality.

This asymmetry is the central problem of product development in the AI era, and it is the reason this book exists.

Research, at its core, is the evaluation function. It is the part of the product system that takes a proposed option and checks it against reality—not the reality of the meeting room or the exec review, but the reality of the person who will actually use the thing, in the context they live in, with their actual goals, constraints, confusion, and the limited time they have.

That evaluation function was designed for a world where option generation was slow and expensive. There was time to evaluate because there were not that many options to evaluate. A team ran a study on the thing they were building because there was one thing, or maybe two, and it took a quarter to build. Now there are dozens of options, they can all be built in a week, and nobody has time for a study on any of them.

There is something else worth naming here, because it is often missed. When option generation is cheap, the perceived value of evaluation drops. Economically, more options should make evaluation more valuable, and that is true. But psychologically, the opposite happens. When generating an option takes thirty seconds and evaluating it takes two days, teams start treating options as disposable. Why evaluate when you can just ship and see what happens? Why study when you can run an experiment? The implicit logic is that the cost of being wrong is lower than the cost of being slow.

And for many decisions, that logic holds. But it breaks down at the moments that matter most: when the wrong default gets baked into the product, when a confused user churns silently, when a feature that tested fine in a two-week experiment slowly erodes trust over six months because nobody understood why users were engaging with it in the first place.

The answer is not to evaluate everything. That was never possible. The answer is to have a system for deciding what to evaluate, how to evaluate it at the right speed, and how to communicate findings so they actually change decisions. That system is what this book builds.

Where AI Actually Touches Research

Before narrowing to the specific territory this book covers, it is worth mapping the full landscape. AI is touching UX research in four distinct areas, and they are not equally mature, useful, or interesting.

The first is recruitment and planning. AI can accelerate participant screening, parse screener responses for consistency, flag mismatches, and assist with scheduling logistics. This is useful and relatively uncontroversial. It makes an administrative process faster. It does not change the nature of the research itself. Most major recruitment platforms are already building this in, and it works well enough that it does not need a book.

The second is analysis and synthesis. This is where most practitioners are using AI today: pasting transcripts into tools like ChatGPT, asking for themes, generating summaries, using AI to code qualitative data, cluster responses, and draft initial reports. The *User Interviews AI in UX Research Report*[1] found that analysis and synthesis is the most common phase for AI use. This is real, growing, and important. But it is also the area where the risks are best understood: overclaiming, smoothing away contradiction, and losing the evidence trail. Chapter 3 covers these failure modes and

1 https://www.userinterviews.com/state-of-user-research-report

their mitigations in detail. Analysis assistance is a tool in the workflow. It is not an operating model.

The third is synthetic users: AI-generated personas that simulate participant responses without involving real people. I have written extensively about this on my blog, and I will be direct here: synthetic users are not research. They are a language model's prediction of what a user might say, based on training data that does not include your users, your product, or your context. They cannot be confused, surprised, or wrong in ways that reveal something you did not expect. They produce confident, plausible, and ultimately ungrounded output. The market for synthetic users is people who want the feeling of research without the inconvenience of actual participants. This book is not for them.

The fourth is AI-moderated research with real participants. This is where a researcher designs a study, writes the questions and probes, and sets guardrails, and an AI tool conducts the sessions with actual people who respond via text, audio, or video, asynchronously and in parallel. The researcher controls the design. The AI handles the execution. Real participants provide real responses that can be audited, traced, and challenged. This is the category that changes what is structurally possible. It is the only category that produces new evidence from real human behavior, and it is the category this book is built on.

The distinction matters because the first three categories improve or automate parts of an existing workflow. They make the current research process faster or cheaper at specific steps. The fourth category creates an entirely new operating mode. It makes it possible to run a credible study with real participants in twenty-four to seventy-two hours—something that was not achievable through any combination of traditional methods, regardless of how fast you moved. That structural shift is what enables the operating model in this book. Without it, the taxonomy collapses. The modes do not exist. The routing logic has nothing to route to.

This book focuses on the fourth category and builds the complete system around it: how to design studies for AI moderation, how to audit the data it produces, how to route questions to the right speed and depth, how to scale the practice across an organization, and how to protect the deep human work that no tool can replace. The other three categories appear where relevant, but they are not the engine. AI-moderated research with real participants is the engine.

What AI-Accelerated Research Unlocks (and What It Cannot Replace)

Before we go further, a boundary that the rest of the book depends on. Because without it, some readers will conclude that this book argues that all research should be fast, that deep work is dead, or that AI has solved everything. That is the opposite of the argument.

AI-accelerated research unlocks a specific capability: the ability to get real user signal on narrow, scoped, near-term product questions in hours or days instead of weeks. That is genuinely useful. It means teams do not have to guess on decisions where a small amount of evidence would meaningfully improve the outcome. It means researchers can serve more decisions with credible signal instead of choosing between one deep study and ten unsupported opinions. It means the gap between decision velocity and evidence velocity can shrink.

But it unlocks this by constraining scope, not by removing constraints. Fast research works when the question is narrow, the domain is understood, the ambiguity is low, and the decision is near-term. When the insight expires in a week, a six-week study is the wrong tool. A focused, well-scoped, fast study—with explicit boundaries on what it can and cannot support—is the right tool. This book teaches you how to build and run that.

What AI-accelerated research cannot replace is everything else. It cannot replace deep research—the work where you sit with users for hours and learn things you did not know you did not know. It cannot replace strategic research that reframes the problem space. It cannot replace longitudinal work that reveals how behavior changes over time. It cannot replace the slow, deeply human work of understanding context, motivation, and the messy reality of how people live with technology. If you try to fast-research your way through a foundational question, you will get a tidy answer to the wrong question and feel productive while doing it.

I also heard the reverse problem in my interviews: teams that skipped fast research entirely because they believed all research had to be deep and rigorous or it was not worth doing. Those teams ended up doing no research at all on the vast majority of their decisions, because deep research on every question is impossible. Perfectionism about method is just as dangerous as recklessness about method. Both lead to the same place: decisions without evidence.

Here is the concern that motivated this book more than anything else. If fast research is the only research that gets done, teams will eventually stop understanding their users at a level deeper than task completion and preference ranking. They will have signal about surface behavior and none about underlying motivation. They will know that users prefer option A over option B but have no idea why—which means they will have no basis for predicting what users will prefer when the options change next quarter. Fast research without deep research is navigating with a flashlight instead of a map: you can see what is immediately in front of you, but you will miss the cliff until you are falling.

The goal is not to replace one tempo with another. The goal is to have both: deep research to build understanding, fast research to inform decisions. Two complementary modes, with clear rules about when to use each and a governance system that prevents the fast one from eating the slow one alive. Because it will try. Stakeholders love fast answers. They will ask for more, then more, then will start suggesting that all research should be this fast, and what were those six-week studies for anyway? A research function needs a system that can absorb that pressure without collapsing. We will build one.

Where This Book Starts

This chapter focused on the shift. The rest of the book focuses on what to do about it.

In Chapter 2, we examine where UX research sits within this new tempo and why its traditional positioning is breaking down. In Chapter 3, we define AI-moderated research clearly: what the tools do well, what they do poorly, and where researcher judgment is the only thing standing between useful signal and confident nonsense. Then we introduce the three operating modes: sprint research for medium-complexity questions over one to two weeks, micro research for narrow questions in twenty-four to seventy-two hours, and deep research—the foundational work that builds the Frame everything else operates within. Without it, fast research answers questions within assumptions that have never been tested. The book then covers the practical systems for scoping, designing, running, and communicating the two fast modes, along with the routing logic for determining which mode fits which question, and the governance required to keep the system balanced and prevent the fast modes from overwhelming the slow one.

The tools changed. The tempo changed. The craft does not have to collapse—but it does have to adapt. Adaptation is not about abandoning what makes research valuable. It is about redeploying it: being faster where fast is appropriate, slower where slow is necessary, and explicit about which is which. It requires systems that prevent either mode from buckling under organizational pressure, stakeholder impatience, or the assumption that AI can do the thinking for you.

Let us get into how.

CHAPTER 2

UXR Was Built for a Slower World

Chapter 1 described the landscape shift: product teams generating options faster than they can evaluate them, decision volume rising, and the gap between decision speed and evidence speed widening. This chapter examines the function that is supposed to close that gap and why, in its current form, it often cannot.

UX research has a positioning problem—not a value problem, not a skills problem, but a positioning problem. The value of understanding users before making product decisions has not decreased. If anything, it has increased, because the number of decisions has multiplied and the cost of compounding small, incorrect bets is real. But the way most research functions are structured, staffed, and embedded in product organizations was designed for a tempo that no longer exists. The operating model is mismatched to the environment in which it operates.

In my interviews, this came up constantly, but it was described differently depending on who was speaking. Researchers described feeling squeezed: too many requests, not enough time, declining influence. Product managers described research as valuable but slow—something they wanted but could not afford to wait for. Design leads described a growing distance between research output and the decisions it was meant to inform. And research leaders described a function that was simultaneously more in demand and less integrated than ever.

All of them were describing the same structural problem from different vantage points.

What stood out was how recently this tension had become acute. Most of the people I spoke with did not describe a slow decline. They described a tipping point, usually within the past eighteen to twenty-four months, when the mismatch between research cadence and product cadence shifted from manageable to actively painful. Before that point, research was somewhat slower than the product cycle but still close enough that it contributed. After that point, the gap widened enough that teams began systematically working around research rather than with it. The tipping point was not a single event; it was the cumulative effect of faster tooling, shorter build cycles, and the organizational expectation that if AI makes everything else faster, research should be faster too.

The Traditional Research Cadence and Why It Worked

Before diagnosing what is broken, it is worth understanding what the traditional model was designed to do and why it worked for as long as it did. This is not nostalgia; it is context. Without understanding what the old model was optimized for, it is difficult to see why simply accelerating the research process does not fix the problem.

The standard embedded research model that scaled across tech from roughly 2012 to 2022 looked something like this: a researcher was assigned to one or more product teams (sometimes called pods or squads). They participated in planning, identified research opportunities aligned with the product road map, scoped studies, recruited participants, conducted sessions, synthesized findings, and delivered readouts. A single study, from intake to readout, typically took two to six weeks, depending on method, complexity, and organizational overhead.

This model was built on several assumptions that were reasonable at the time. Product cycles were measured in months or quarters. Features took weeks or months to ship. There was a natural window between idea and commitment where research could contribute. Decisions were relatively large, meaning each one carried enough weight to justify a dedicated study. And the expectation was that research would be thorough, nuanced, and defensible—not fast.

Within those constraints, the model worked. Researchers built deep relationships with their teams. They developed domain expertise and produced work that shaped strategy as well as tactics. They were trusted not because they were fast but because they were rigorous. The readout was an event: people showed up, findings were discussed and debated, and sometimes they were acted on.

Several senior researchers I interviewed described this era fondly but honestly. One researcher with fifteen years of experience at multiple large tech companies said the embedded model was "the best version of what research could be when the pace allowed it." but added that she does not think that pace is coming back. Another described it as "a golden period where the org chart and the product cycle happened to align in research's favor." The point is not that the old model was flawed. The point is that the conditions it depended on have changed.

Those conditions were already eroding before AI entered the picture. Agile shortened cycles. Continuous delivery compressed timelines. Feature flags made it possible to ship and iterate without waiting for a full release. The research window had been shrinking for years. AI did not create the mismatch; it made it impossible to ignore.

What Changed and Why Speeding Up Is Not Enough

The instinctive response to a speed mismatch is to go faster: run shorter studies, recruit faster, synthesize in a day instead of a week. Use AI tools to compress transcription, summarization, and analysis. This is directionally correct, but it is incomplete. And if it is the only response, it creates new problems without solving the underlying one.

Here is why. The traditional research model is not just slow in execution; it is slow by design. The entire workflow assumes a sequential, gated process: scope, recruit, run, synthesize, deliver. Each step depends on the previous one. Each step has its own lead time. Even if you compress each step significantly, the total end-to-end timeline still spans days or weeks, because the steps remain sequential and each involves coordination with other humans.

Speeding up each step helps, but it does not change the fundamental architecture. It is like widening lanes on a highway that has too many traffic lights: each segment is a little faster, but you are still stopping at every intersection. What product teams actually need is a different kind of road for a different kind of trip.

Multiple researchers I interviewed described a version of this frustration clearly. They had adopted AI tools for transcription and initial synthesis. They had shortened their timelines significantly, and they were still being told they were too slow. One researcher at a growth-stage marketplace company said she had cut her average study timeline from three weeks to eight days, and her PM still asked if she could do it in two. "The goalposts moved," she said. "And they will keep moving, because the product cycle keeps compressing. Shaving days off of the old process is not going to get us there."

The issue is not just speed; it is responsiveness. Product teams do not always need a full study. Sometimes they need a quick answer. Sometimes they need a sanity check. Sometimes they need someone to tell them their question is unanswerable with a quick study and requires a different method or a different conversation entirely. The traditional model treats everything as a study. That is both its strength and its limitation. When the only tool in the system is a scoped research project, every question either becomes a project or goes unanswered.

A PM at a developer tools company described this gap precisely. "Eighty percent of the time, what I actually need is not a study. I need a researcher to look at what we are building, tell me if the question I am asking is answerable, and if it is, help me get a directional answer in a few days or a week max. The other 20 percent of the time, I genuinely need a full study. But the research team only offers me one speed, and it is the full-study speed. So I get great research on 20 percent of my questions and nothing on the other 80 percent." He was not frustrated with the researchers. He was frustrated with the system.

A research operations lead at a large e-commerce company described the intake queue her team maintained. Every request went into the same pipeline, and every request was scoped, estimated, and prioritized. The average turnaround was three weeks. She said the system was fair and orderly—and it was also irrelevant to about half the requests that came in, because by the time a study was scoped and scheduled, the team had already made the decision and moved on. "We were running a research factory," she said, "and our customers had already switched to a different supplier." That supplier, in most cases, was intuition.

The Growing Gap Between Research and Decisions

To understand the positioning problem clearly, it helps to look at where research actually sits in the decision-making timeline of a modern product team—not where it is supposed to sit, but where it actually sits.

In the traditional model, research occupies a position between problem identification and solution commitment. The team identifies a question or opportunity, research investigates it, and the findings inform the direction. This works when the gap between identification and commitment is measured in weeks. It breaks when that gap is measured in days or hours.

What I heard consistently in my interviews is that research has drifted from a pre-decision function to a post-decision validation function—or worse, a post-launch rationalization function. Teams make the decision, ship the thing, and then ask research to confirm it was the right call. Or they do not ask at all. Research is not excluded maliciously. It is excluded temporally. The decision window opens and closes before research can contribute.

A product director at a consumer social company described his team's relationship with research as "genuinely respectful but logistically broken." He said the team valued the researcher's judgment, referenced past studies regularly, and wanted to involve research in more decisions. But the cadence did not line up. "By the time research can deliver, we have already committed. Not because we do not care but because the timeline forces the choice. Wait for evidence or ship on schedule. We ship on schedule almost every time."

This is not a failure of trust. It is a failure of the operating model. The research function is structured to produce high-quality, comprehensive answers on a timeline that no longer matches how decisions get made. The result is that research gets used selectively, on the decisions large enough or uncertain enough to justify the wait. Everything else gets decided without it.

The math is not complicated. If a research team can produce maybe two to four studies per researcher per quarter, and the product teams it supports are making dozens of decisions per sprint, the coverage rate is in the single digits. Most decisions will not have research input. That is simple arithmetic. The capacity of the function does not match the demand of the environment.

What makes this worse is that the decisions research does cover tend to be the large, strategic ones. Those are important, but the accumulation of small decisions that go uninformed is where the real erosion happens. Nobody ships a catastrophically bad product because they got one big decision wrong. They ship a mediocre product because they got fifty small decisions slightly wrong, and those slight misalignments compounded into an experience that feels off without anyone being able to say exactly why.

Several research leaders I spoke with were candid about this. One described it as "the coverage illusion." Her team ran high-quality studies that influenced important decisions. But those studies touched maybe 10 percent of the product decisions made in a quarter. The other 90 percent were made based on opinion, analytics, or nothing. "Leadership sees the studies we do and thinks research is embedded," she said. "They do not see the 90 percent we never touch."

What Happens When Research Cannot Keep Up

Chapter 1 described the gap-fillers: opinion dressed as insight, analytics misread as understanding, AI-generated insight theater, and decision debt. Here I want to talk about something different—not what fills the gap, but what happens to the research function itself when it consistently cannot keep up.

Three patterns emerged from my interviews, and they tend to compound.

The first is marginalization through respect. This sounds paradoxical, but it is common. The research team is genuinely valued. Leadership says the right things. The researcher is invited to meetings. But the function gradually shifts from pre-decision to post-decision. Research becomes the team that confirms what was already decided rather than the team that shapes what gets decided. Studies get scoped around decisions that have already been made. Readouts present findings that align with directions already committed to. The researcher becomes a validator, not an investigator. This is comfortable in the short term and corrosive in the long term, because it trains the organization to see research as a rubber stamp rather than a source of genuine signal.

The mechanism is subtle. Nobody sends an email saying "we have already decided, please confirm." Instead, the brief arrives preloaded with assumptions. The research question is framed as "can you confirm (aka validate) that users understand this flow?" rather than "determine whether this flow works for users and why." The difference looks small on paper. It is enormous in practice. A validation frame tells the researcher what to find. An investigation frame tells the researcher what to look for. Teams that have slid into validation framing often do not realize it until someone points it out, and even then, the organizational incentive to confirm rather than challenge is strong enough that pointing it out does not always change the behavior.

A researcher at a large enterprise software company described realizing this had happened to her when she noticed that every study her team ran in Q3 was framed as "validate the direction." Not investigate. Not explore. Validate. The word choice was not accidental. The teams had already committed. They wanted confirmation, not insight. She said the studies still produced useful findings, sometimes findings that contradicted the direction. But the organizational posture was already set. Contradictory findings were received as interesting footnotes, not decision-changing evidence.

The second pattern is the retreat into craft. When a research function feels its influence shrinking, it sometimes responds by doubling down on methodological rigor. The studies become more thorough, more carefully designed, more defensible. The discussion guides get longer. The sample sizes are considered. The synthesis becomes more nuanced. All of which is genuinely good work—and all of which makes the function slower and less responsive at exactly the moment it needs to be faster.

This is not irrationality. It is a rational response to a threatening environment. If the value of research is being questioned, proving that the work is rigorous feels like the right defense. The problem is that rigor, as traditionally defined in UXR, is measured by thoroughness, not by decision impact. A perfectly rigorous study that lands after the decision was made has zero impact regardless of its methodological quality. Several research managers I interviewed described watching their teams respond to organizational pressure by going deeper instead of going faster, and recognizing too late that depth was not what the organization was asking for.

There is a version of this that is especially common among researchers who came up through academic or agency backgrounds, where rigor was the primary currency. In those environments, a well-designed study was an end in itself. The quality of the work was the deliverable. In a product organization, the quality of the work is a means to an end, and the end is a better decision. When researchers treat methodological rigor as the goal rather than the instrument, they optimize for the wrong thing—not because rigor is unimportant, but because rigor that does not connect to a decision is rigor that does not matter to the organization paying for it. This is a hard pill for many researchers to swallow, and it came up frequently in our conversations. The ones who had made peace with it described it as the most important shift in their professional development.

The third pattern is quiet irrelevance. This is the endgame of the first two patterns and the hardest to see from the inside.

The research team still exists. It still runs studies. It still produces readouts. But the product organization has learned to operate without it on most decisions. Research is consulted on large bets and ignored on everything else. The team is not cut. It is not reorganized. It is just gradually routed around. Decisions flow around research the way water flows around a rock. The rock is still there. The water does not care.

The most insidious version of this pattern is when the research team's own metrics look healthy. Studies completed per quarter: on target. Stakeholder satisfaction scores: high. Readout attendance: good. The team feels productive. But if you look at the decisions being made across the product organization and trace which ones had research input, the coverage is thin. The team is busy doing good work on a small number of questions while the majority of product decisions are made without any research at all. Activity looks like impact. It is not.

A VP of product at a midsize B2B company described this state with uncomfortable clarity. "The research team does great work," he said. "But we have basically two tracks. There is the research track, where a study happens and findings come back in three weeks. And there is the product track, where decisions happen in real time. The two tracks run in parallel and occasionally intersect. They should be the same track. They are not."

If any of this sounds familiar, it is not because your research team is failing. It is because the operating model was designed for a different environment, and the environment changed. Recognizing that is the first step toward fixing it.

The Identity Question: Service, Practice, or Strategic Function?

Underneath the operational mismatch is a deeper question that most research functions have not resolved, and it matters because the answer determines what adaptation looks like.

Is UX research a service, a practice, or a strategic function?

If it is a service, then research exists to answer questions that product teams bring to it. The value is in the answers. The quality metric is customer satisfaction: did the team get what it needed when it needed it? The speed problem is a service delivery problem, and the fix is operational: faster intake, faster execution, better tooling, more efficient workflows.

If it is a practice, then research exists to maintain a standard of evidence and inquiry within the product organization. The value is in the discipline. The quality metric is methodological integrity: are the right questions being asked, with the right methods, and are conclusions appropriately bounded? The speed problem is a scope problem, and the fix is better triage: not everything needs a study, not every study needs to be comprehensive, and the practice should define what "good enough" looks like at different levels of speed and risk.

If it is a strategic function, then research exists to shape how the organization understands users and makes decisions. The value is in influence. The quality metric is decision quality: are better decisions being made because research exists? The speed problem is a positioning problem, and the fix is structural: research needs to be embedded in the decision-making process itself, not adjacent to it.

Most research functions are a blend of all three, which is fine in theory and messy in practice. The problem is that different stakeholders hold different mental models of what research is, and those models create conflicting expectations. A PM who sees research as a service wants fast answers. A research director who sees research as a practice wants methodological integrity. A VP who sees research as a strategic function wants influence on the product road map. All three are legitimate. All three create different pressures. And when the tempo shifts and resources are constrained, the unresolved tension between these identities becomes the fault line along which the function cracks.

In my interviews, I asked researchers and research leaders to describe how their organization thought about the research function. The most common answer was some version of "it depends on who you ask." PMs generally treated research as a service. Researchers themselves generally identified as practitioners. Executives, when they thought about it at all, talked about research in strategic terms during planning and service terms during execution. The disconnect was rarely surfaced explicitly. It just manifested as friction: researchers feeling undervalued, PMs feeling underserved, and leadership feeling like research was important but somehow never quite aligned with the pace of the business.

A research director at a large consumer technology company described the tension this way. "In planning, we get treated as strategic. We are in the room. We shape the road map. In execution, we get treated as a service desk. The same VP who asks us to define the research agenda in January is asking us to turn around a quick study in two days in March. Both of those things are reasonable. But they require completely different operating models, and we are trying to do both with the same team, the same process, and the same expectations."

The honest answer to the identity question is that research needs to be all three at different times, for different questions, with clear systems for deciding which mode to operate in. A strategic function that cannot deliver fast answers when they are needed will be routed around. A service that never does deep work will lose its ability to shape decisions at a strategic level. A practice that ignores organizational tempo will produce beautiful work that nobody uses.

What this means in practice is that the research function has to stop expecting the organization to adapt to its preferred way of working and start adapting its way of working to the organization's reality. A doctor does not insist that every patient needs an MRI because MRIs produce the best images. They match the diagnostic tool to the clinical question. Research needs to do the same. Some questions need an MRI. Some questions need a quick check. The skill is in knowing which is which and being credible at both.

This is not a comfortable answer. It means the research function needs to operate in multiple modes simultaneously, and it needs systems for switching between them. It means saying yes to some things and no to others, not based on capacity alone but on a deliberate judgment about what kind of question is being asked and what kind of answer it requires. It means having governance that protects deep work from being consumed by fast work, and quality standards that prevent fast work from becoming sloppy work.

Most research functions do not have these systems. They have a single pipeline that treats every request approximately the same way. That pipeline was serviceable when the tempo was slower and the volume was lower. Under current conditions, a single pipeline cannot serve multiple operating modes. It will default to whichever mode gets the most organizational pressure, which is almost always speed, and the other modes will atrophy.

One of the clearest examples came from a research manager at a fintech company who described what happened when her team started responding to the pressure for speed without changing its operating model. The team shortened its studies, cut synthesis time, reduced sample sizes, and tried to run the same process faster. Within a quarter, three things happened. Quality dropped visibly, to the point where a PM pushed back on a readout because the findings contradicted what he was seeing in analytics and the sample was too small to be credible. Researcher morale declined because the work felt rushed and unsatisfying. And the team was still too slow for the fastest decisions, because a compressed version of the old process was still a process that took a week. They had traded quality for speed and gotten neither. That is what happens when you try to solve a structural problem with execution pressure.

What the New Positioning Looks Like

If the old positioning was "research is the function that runs studies," the new positioning needs to be something closer to "research is the function that ensures decisions are informed by evidence at the appropriate level of rigor and speed." That is a broader mandate and a harder one. It requires researchers to be comfortable operating across a wider range of methods, timelines, and levels of certainty than the traditional model demanded.

Concretely, this means research needs to operate in at least three modes.

The first is deep research. This is the traditional mode: foundational studies, strategic investigations, longitudinal work. It takes weeks or months. It produces understanding that shapes the product direction. It cannot be rushed without destroying its value. Every research function needs to protect time and budget for this work, because it is the foundation on which everything else rests.

The second is sprint research. This is the mode for questions that have moderate ambiguity and moderate stakes, where a few days to a couple of weeks of focused investigation can produce evidence that meaningfully changes direction. Sprint research is faster than traditional studies but still involves researcher judgment throughout. It requires tighter scoping and faster synthesis, but it is not fundamentally different in kind from deep research. It is different in scope and timeline.

The third is micro research. This is the mode for narrow, specific, near-term product questions where the decision is clear, the ambiguity is low, and the question can be answered with real participants in twenty-four to seventy-two hours. Micro research is not a shortcut. It is a distinct operating mode with its own quality standards, its own scoping discipline, and its own rules for what you can and cannot conclude.

We will define sprint research and micro research in detail in later chapters. For now, the point is structural. A research function that can only operate in one mode will be either too slow for most decisions or too shallow for the decisions that matter. The new positioning requires multiple modes, with clear criteria for routing questions to the right one, and governance that prevents the fastest mode from cannibalizing everything else.

In my interviews, the teams that were navigating the tempo shift most effectively were the ones that had, in some form, built multi-modal research operations. They did not all use the same language. Some called it tiered research. Some called it a research menu. One team literally had a service-level matrix with three columns: deep, fast, and self-serve. The specifics varied, but the principle was the same: not every question gets the same process, and the function is responsible for matching questions to the right mode, not just executing the mode it is most comfortable with.

A research lead at a health technology company described the shift in mindset this required. "The hardest part was not learning new methods or tools," she said. "The hardest part was letting go of the idea that every question deserves a full study. Some questions deserve a full study. Some questions deserve a focused weeklong sprint. Some questions deserve a gut check with seven users and an honest statement of what you can and cannot conclude from that. The craft is in knowing which is which, not in treating everything the same."

That judgment—the ability to look at a question and route it to the right mode at the right speed with the right level of rigor—is the core skill of research in the current environment. The rest of this book is about building the systems that make that judgment repeatable, teachable, and organizationally sustainable.

The Tool Adoption Trap

The tech industry has spent more than $700 billion on AI. A 2026 NBER study[2] asked 6,000 executives across four countries whether any of it moved the needle. Eighty-nine percent reported no productivity boost. A separate study[3] of experienced developers using the latest AI tools found they completed tasks 19 percent slower with AI than without it. These findings show what happens when you adopt tools without changing how the function is organized, governed, and run. The tools compress individual tasks. The operating model determines whether that compression produces anything useful.

Seven hundred billion dollars. Nineteen percent slower.

This is not a new pattern. Economists have been writing about it since the 1980s, when computers showed up everywhere except in the productivity data. Robert Solow captured it in 1987: you can see the computer age everywhere except in the productivity statistics. Forty years later, Apollo's chief economist, Torsten Slok, applied the same observation to AI.[4] The pattern is so consistent it has its own name: the productivity paradox. Different tool, same paradox, same reason.

The reason is always the same: the tool is not the problem; the structure the tool was dropped into is the problem.

This is not the first time the industry has been here. Remember agile? The promise was faster delivery, empowered teams, fewer approvals, better software. What happened in most organizations

2 https://www.nber.org/papers/w34836

3 https://metr.org/blog/2025-07-10-early-2025-ai-experienced-os-dev-study/

4 https://www.apolloacademy.com/waiting-for-the-ai-j-curve/

is that they renamed their meetings, bought Jira licenses, hired a Scrum Master who looked increasingly haunted as the months went by, and kept doing exactly what they were doing before. The same decisions were made by the same people in the same way—just with more standups.

Then design thinking arrived in a cloud of sticky notes and empathy maps. Innovation labs opened. People started saying "how might we" in meetings where the answer had already been decided. The org chart stayed identical. The VP who ignored user research before design thinking ignored it after. The sticky notes went up and came down, and nothing changed except that everyone owned a marker.

Then lean arrived, then OKRs, and then digital transformation, which was agile with a larger consulting invoice.

Now it's AI—same play, bigger budget, much better graphics.

None of these approaches were wrong. Agile, done well, genuinely produces better software faster. Design thinking, done well, genuinely produces more user-centered products. The problem was never the idea. It was that organizations adopted the label before building the conditions that make the idea work. The framework arrived years after the tools. By then, most organizations had already decided the transformation had failed and moved on to the next one.

The organizations that are actually getting productivity gains from AI are not the ones that bought the best tools. The 2024 BCG[5] study and the 2025 P&G[6] study both point to the same finding: the organizations showing real gains are the ones that redesigned the

---

5 https://www.bcg.com/publications/2024/gen-ai-increases-productivity-and-expands-capabilities
6 https://papers.ssrn.com/sol3/papers.cfm?abstract_id=5188231&

work around the tools. They started from the question of what they were trying to produce and worked backward to what the operating model needed to look like to produce it, with AI as a design constraint rather than an afterthought. That is not a tool decision. That is an organizational design decision.

In research specifically, the failure mode is almost poetic in its consistency. Transcription got automated. Synthesis got faster. Repositories started filling up more quickly. And then the research function continued operating exactly as it always had: answering questions as they came in from whoever asked the loudest, running studies on timelines that had nothing to do with when decisions were actually being made, producing decks that got acknowledged in meetings and then quietly ignored.

The service model is still the service model. AI made it faster. A faster service model is still a service model. You gave someone a faster car and left them in the same traffic jam. The car is not the problem.

This is the problem the rest of this book is trying to solve: not which tools to use, but how to build the operating model that makes the tools mean something. The two are not the same question. Most organizations are only asking the first one.

Where This Goes Next

This chapter described where research sits today in most organizations and why that positioning is under strain. The operating model was built for a slower world. The tempo changed. Research did not adapt structurally, even when individual researchers adopted faster tools and methods. The result is a function that is valued but underutilized, respected but routed around, and stuck between identities that pull in different directions.

None of this is inevitable. The research functions that are navigating this shift successfully share a common trait: they stopped trying to run the old model faster and started building a new model with multiple speeds. They defined what kinds of questions get what kinds of treatment. They built routing systems so that fast questions get fast answers and deep questions get the time they need. They set governance so that speed does not cannibalize depth. And critically, the fastest modes in their systems were not traditional research compressed into shorter timelines. They were built on a category of AI moderation tooling that made structurally new research modes possible. Without those tools, the multimodal operating model is an aspiration. With them, it is buildable. These are not personality traits or organizational luck. They are systems, and this book teaches you how to construct them.

The fix is not to go faster within the old model. The fix is to build a new operating model with multiple modes, clear routing, and governance that protects the integrity of each mode. Chapter 3 begins that work by defining what AI-moderated research actually means in practice: what the tools do, what they do not do, and where the researcher's judgment remains the only thing standing between useful evidence and confident noise.

CHAPTER 3

AI-Moderated Research as a Response

The previous two chapters described a structural problem: product teams making decisions faster than research can inform them, and a research function whose operating model was built for a tempo that no longer exists. Chapter 1 mapped where AI touches research and identified AI-moderated interviews with real participants as the capability that changes what is structurally possible. This chapter goes deep on that capability—not the hype version of it, not the vendor pitch, but the honest, practitioner-level assessment of what AI-moderated research actually does, where it breaks, and why, despite real limitations, the value is significant enough to build a new operating model around.

This chapter is not an endorsement of any specific tool. It is also not a warning label. It is an attempt to be precise about capabilities, limitations, and the judgment required to use these tools responsibly. Because the discourse around AI in research tends to split into two camps: people who think AI will automate research entirely and people who think AI has no place in qualitative work at all. Both camps are wrong, and both are loud enough to drown out the practitioners in the middle who are actually using these tools, finding real value, encountering real problems, and figuring out how to make it work.

The automation camp tends to be populated by tool vendors, executives who want to reduce headcount, and people who have never actually run a research study but are confident that AI can handle it. Their argument rests on a misunderstanding of what research is. They think research is data collection. It is not. Data collection is one step. The hard parts, the parts that determine whether the output is credible and useful, are question design, quality control, interpretation, and communication. Those require human judgment that current AI cannot provide.

The rejection camp tends to be populated by researchers who see any AI involvement as a threat to rigor, craft, or their jobs. Their argument rests on a comparison to an ideal that rarely exists in practice. They compare AI-assisted research to a perfectly designed, perfectly moderated, perfectly synthesized traditional study. That comparison is valid in theory. In practice, the alternative to AI-assisted research is usually not a perfect study. It is no study at all, because the team did not have three weeks and made the decision without evidence.

In my interviews, the practitioners in the middle were the majority. They were not ideological about AI. They were pragmatic. They had adopted tools, hit limitations, developed workarounds, and formed opinions grounded in experience rather than theory. This chapter draws heavily on what they told me.

What the Tools Actually Do

Chapter 1 mapped four areas where AI touches research and identified AI-moderated interviews with real participants as the capability this book is built on. The other three—recruitment screening, analysis and synthesis, and synthetic users—are addressed where relevant throughout the book. This section goes deeper on what AI moderation actually looks like in practice, because the rest of the operating model depends on understanding it precisely.

Tools like Outset allow a researcher to design a study, write the questions and probes, define the guardrails, and then have the AI conduct the interviews asynchronously with recruited participants. The participants respond via text, audio, or video. The AI follows the researcher's script, asks follow-up probes based on participant responses, and captures everything for review. The researcher designs the study and interprets the results. The AI handles the moderation at scale.

This is not a small thing. In a traditional workflow, moderation is the most time-constrained step. A researcher can only be in one session at a time. Scheduling is a coordination challenge: time zones, availability, no-shows, rescheduling. AI-moderated interviews remove that bottleneck entirely. A study that would take a week to schedule and run with five participants can be launched in a morning and have responses back by the next day—with ten participants, or twenty. The constraint shifts from "how many sessions can I physically conduct" to "how many responses can I meaningfully analyze."

Several researchers in my interviews described this as the single biggest practical change in their workflow—not synthesis acceleration, not transcription, but the moderation step. One researcher at a consumer technology company said she used to spend roughly 60 percent of her study time on logistics and moderation. With AI-moderated interviews, that dropped to maybe 15 percent, and the freed time went into better question design and more careful interpretation. That is a significant reallocation of where researcher time goes.

The tools will continue to improve. Moderation will handle more complex follow-up logic. Transcription and summarization will get more nuanced. Recruitment screening will get better at detecting fraud. The point is not to wait for the tools to be perfect. The point is to use them where they are good enough today and build the judgment and quality systems that keep the work credible regardless of how the tools evolve.

What Changes When the Moderator Is Not Human

What actually changes when the moderator is not human is worth understanding precisely, because it determines when AI moderation is the right instrument and when it is not.

The traditional research interview is a social interaction: two humans in a room or on a video call, navigating a conversation with all the unspoken rules that come with being human in the presence of another human. The participant reads the moderator's face. They pick up on microexpressions, nods, pauses, shifts in posture. They calibrate their answers based on what they think the moderator wants to hear, or what they think will make them look competent, or what feels socially appropriate to say out loud to a stranger. The moderator, in turn, reads the participant. They notice hesitation, confusion, enthusiasm, discomfort. They adjust in real time. They sit in silence when the participant needs to think. They lean in when something unexpected surfaces. They make a judgment call every few seconds about where to push and where to let go.

That dance produces rich data. It also produces biased data. The social dynamics of a human-to-human interview are both its greatest strength and its most persistent confound.

AI moderation removes the dance entirely, and the effects of that removal run in multiple directions at once.

Participants perform less. There is decades of research showing that self-administered and computer-mediated modes produce less socially desirable responses than interviewer-administered ones. People report more honestly about sensitive topics when a human is not in the room. The emerging research on AI-moderated interviews specifically confirms that this pattern holds. A large-scale study from researchers at the London School of Economics and Harvard found that roughly half of respondents preferred interacting with an AI interviewer over a human, and the primary reason was that they perceived the AI as a nonjudgmental entity.[7] A separate study on chatbot-based data collection for HCI research found the same pattern: participants preferred chatbots for sharing personal thoughts specifically because of the absence of social judgment.[8] And a 2024 study by Nesta reported that participants

7 https://blogs.lse.ac.uk/impactofsocialsciences/2024/10/30/ai-can-carry-out-qualitative-research-at-unprecedented-scale/

8 https://dl.acm.org/doi/full/10.1145/3717511.3747079

in AI-led interviews felt less pressure and judgment compared to conventional research interviews.[9]

For product research, this matters. When a human moderator asks, "What do you think of this design?", participants feel social pressure to be polite, constructive, and articulate. They give the kind of answer they would give to a colleague. When an AI asks the same question, that pressure drops. Several researchers in my interviews described getting harsher, more direct feedback from AI-moderated sessions than from equivalent human-moderated ones. One researcher at a consumer fintech company described running the same study design in both modalities and being surprised by the difference. "The human-moderated participants were diplomatic," she said. "The AI-moderated participants were honest. Sometimes brutally so."

This is not universally positive. Diplomatic answers carry nuance. A participant who says "I think this could work, but I had to read it twice" is telling you something different from a participant who says "this is confusing." Both are useful, but the first gives you more to work with. Still, for research questions where you want unfiltered reactions—especially around comprehension, friction, and objection mining—the reduced social performance in AI-moderated sessions can produce cleaner signal.

Sensitive topics get more honest treatment. This follows directly from the reduced social desirability effect, and both the evidence discussed above and practitioner experience support it. When the moderator is not human, the judgment anxiety decreases. Participants are not worried about what the AI thinks of them. They

9 https://www.nesta.org.uk/blog/can-ai-drive-innovation-in-qualitative-research-heres-what-weve-learnt/

are not managing their self-presentation in the same way. A researcher working in health technology described running studies on medication adherence where AI-moderated participants were significantly more forthcoming about skipping doses and the reasons why. The human-moderated sessions produced more socially acceptable answers. The AI-moderated sessions produced more truthful ones.

This effect has real implications for research design. If your question touches anything participants might feel embarrassed, judged, or exposed about, AI moderation may actually produce better data than human moderation. That is counterintuitive for researchers trained to believe that rapport is the key to disclosure. In some cases, the absence of rapport is what unlocks honesty.

Rapport is real, but its effect on data quality is more complicated than assumed. A 2026 biometric study comparing AI and human interviewers measured both subjective and physiological responses to AI versus human interviewers. Participants reported a meaningfully stronger sense of connection with human interviewers and gave them higher overall evaluations. That is not surprising. What is surprising is the second finding: that stronger emotional rapport did not translate into better data collection or disclosure. Participants were equally willing to share personal information regardless of whether they spoke to an AI or a human. The rapport advantage was real as an experience but did not produce measurably different data.

This finding matters for how researchers think about AI moderation. The instinct is that rapport equals better data. The emerging evidence suggests that rapport equals a better experience for the participant, which is valuable but is different from data quality. Researchers who resist AI moderation because "you cannot build rapport with a machine" may be correct about the rapport and wrong about the implication.

Depth is the trade-off, and the research and practitioner experience align on this clearly. The same absence of human connection that reduces performance and unlocks honesty also limits how far the conversation goes. A skilled human moderator builds rapport not as an end in itself but as a tool for getting participants to go deeper. The participant who trusts the moderator will share the second layer of reasoning—the story behind the answer, the context that explains the behavior, the thing they would not say to a machine because the machine did not earn it.

The Nesta study was more direct: the researchers found that one human-led interview could surface as many or more valuable insights as dozens of AI-led interviews and that AI-powered interviews miss the subtle, contextual richness that comes from skilled human interviewers using follow-up prompts and reading nonverbal cues. That gap is real and the honest position is to name it clearly. It is also narrowing. Some platforms are beginning to incorporate visual intelligence capabilities—reading facial expressions, hesitation, and emotional response in video sessions—in ways that start to approximate what a skilled human moderator notices. The technology is early and not yet a substitute for human judgment in depth interviews, but the trajectory matters: the limitations described here reflect the current state of the tooling, not a permanent ceiling.

Researchers in my interviews described this as the "answer but don't tell" pattern. Participants answer what they are asked with reasonable completeness. They do not tell you the thing they were not asked about but that matters more than what you were asking. In human-moderated sessions, those moments emerge through rapport, through comfortable silences, through the moderator sensing that the participant has more to say and gently holding space for it. AI moderators do not hold space. They move to the next question.

This is not a fatal limitation. It is a design consideration. For research questions where the answer is the data (comprehension, preference, friction identification), the shallower interaction is not a significant loss. For research questions where the context around the answer is the data (motivation, mental models, decision-making processes), the shallower interaction is a real cost.

AI moderation also produces something human moderation cannot: a consistent baseline across participants. A human moderator's energy, phrasing, and follow-up instincts vary across sessions in ways that introduce noise into the data. The participant who gets a tired moderator on session eleven gets a different experience than the one who got an energized moderator on session two. AI moderation removes that variability. Every participant receives the same prompts, the same probes, and the same pacing. The data is shallower per participant but more comparable across participants, which matters when you are trying to identify patterns rather than explore individual experiences in depth.

Response volume is higher, but that is not the same as richness. The LSE study found that participants in AI-moderated interviews wrote significantly more words compared to open-text survey fields. That is encouraging if the comparison is surveys. It is less encouraging if the comparison is a skilled human interviewer who can pull three paragraphs of context from a single follow-up question. More words does not mean more insight. It means more data to process. And the researcher needs to assess whether that additional volume contains additional signal or just additional noise.

Who participates well may not be random. This is my own observation from running AI-moderated studies and from what researchers described in my interviews, rather than a research finding, but it came up consistently enough to flag. Not every participant interacts with AI the same way. Some are comfortable with the format and give rich, thoughtful responses. Others treat it like a survey and give minimal answers to get through it. The variability in how participants engage with an AI moderator is higher than with a skilled human moderator, because a good human moderator adapts to each participant and draws them out. The AI treats everyone the same.

This creates a subtle selection effect in your data. The participants who engage well with AI moderation are overrepresented in the useful responses. Those participants who disengage are underrepresented, not because they were screened out but because their thin responses carry less weight in synthesis. Several researchers described noticing this pattern only after comparing their AI-moderated data across multiple studies. The participants who gave the richest responses to AI tended to share certain characteristics: comfort with technology, willingness to type or speak at length without prompting, and a disposition toward giving detailed answers even without social incentive. That is not a representative sample of most user bases. It is a subset, and the researcher needs to be aware of it. Formal research on this specific selection effect in AI-moderated product research has not been published yet, but it is worth watching for as the field matures.

The net effect is not better or worse; it is different. AI-moderated interviews produce data that is more honest on sensitive topics, less performative, more blunt, more uniform in format, and shallower in contextual richness. Human-moderated interviews produce data that is more nuanced, more narrative, more responsive to the individual participant, and more susceptible to social desirability bias. Neither is objectively superior. They produce different kinds of data, and the researcher needs to understand the difference to design studies that play to the strengths of each.

This is why the operating model in this book has two modes that use AI moderation differently. Micro research leans into the strengths of AI moderation: fast, parallel, direct responses to concrete questions where depth is less important than coverage and honesty. Sprint research uses AI moderation for the first layer and then brings in human follow-ups for the second layer, precisely because the depth and contextual richness require a human in the conversation.

The common mistake is treating AI moderation as a cheaper version of human moderation. It is not a cheaper version. It is a different instrument. Using it well means understanding what it captures that human moderation does not, what it misses that human moderation catches, and designing your study so the instrument matches the question. A researcher who understands these dynamics will produce better work with AI tools than a researcher who treats them as a simple substitute for being in the room.

Where the Tools Break

Now the honest part. Because if this chapter only described what the tools do well, it would be a brochure, not a book.

Before we dive in, let's talk briefly about framing. Most of the failure modes described in this section also exist in human-moderated research. Less experienced moderators lead participants. Synthesis overclaims. Sessions go off-script. Evidence trails get thin. The rejection camp described at the start of this chapter compares AI moderation to an ideal that rarely exists in practice. This section should not inadvertently do the same thing. The difference with AI moderation is not that these problems are new. It is that they become more visible and more auditable. Every follow-up the AI asked is in the transcript. Every probe is reviewable. The failure modes are catchable in a way that a human moderator's subtle leading question—delivered and forgotten in the flow of a live session—often is not. The quality gates described below work precisely because the evidence is there to audit. That is not a weakness of AI moderation. It is, handled correctly, one of its underappreciated strengths.

The tools have real limitations that, if ignored, produce work that looks like research but is not. These are common failure modes that showed up repeatedly in my interviews and in direct experience with the tools. They need to be named clearly, because the risk is not that these tools are useless. The risk is that they are useful enough to create false confidence.

The first and most significant failure mode is leading and shaping by the AI moderator. When an AI conducts an interview, it follows a guide programmed by the researcher and probes based on participant responses. But the way it phrases follow-ups, the way it sequences questions, and the way it acknowledges responses can subtly lead participants toward certain answers. A human moderator notices when a participant seems confused and can adjust. A human moderator reads body language, hesitation, and tone. A human moderator knows when to sit in silence and let the participant think. AI moderators do not do any of this reliably. They tend to fill silences, rephrase in ways that narrow the response space, and acknowledge answers in ways that signal approval.

The effect is not dramatic. It is not that the AI asks blatantly leading questions (although poorly designed prompts can produce exactly that). It is that the interaction subtly shapes responses in ways that reduce variability and push toward the expected. This is distinct from the social desirability effect described earlier. Participants talking to AI moderators are more blunt and less performative in their overall orientation. But they are also more responsive to the specific framing of each question, precisely because there is no human relationship to push back against. A poorly designed probe does not get questioned or reinterpreted the way it might in a human-moderated session, where the participant feels comfortable saying "I'm not sure what you mean." It gets answered in the direction the phrasing suggests. The result is data that can look clean and convergent while actually reflecting the structure of the prompts more than the reality of the participants. One researcher who had been running AI-moderated studies for about eight months described noticing this pattern when she ran the same study design twice: once with AI moderation and once with researcher-moderated sessions. The AI-moderated data was tidier—clearer themes, less contradiction. At first she thought that meant the AI was more consistent. Then she compared the transcripts and realized the AI was producing less contradiction because it was subtly narrowing the response space. The human-moderated sessions had more mess, more tangents, more unexpected responses, and more signal. Clean data is not the same as good data. The mess is where the insight lives.

The most effective mitigation, based both on what practitioners described in my interviews and on my own experience, is rigorous prompt design and the first-three-participant audit. Design the AI moderator's prompts with the same care you would use for designing a discussion guide. Test them with internal participants before launching. Then, when the study is live, review the first three transcripts in full before letting the rest of the data collect. Read the AI's actual follow-ups. Check for leading phrasing. Check for patterns where participants converge on similar language, which can indicate the AI is shaping rather than probing. If the first three transcripts show leading patterns, stop the study and revise the prompts. This is not optional. It is the quality gate that makes AI-moderated research credible.

The second failure mode is overclaiming from AI-generated summaries. When a tool summarizes interview transcripts, it produces confident, well-structured prose. That prose sounds like findings. It looks like synthesis. It is neither. It is a compression of what was said, organized by pattern and frequency, but without the contextual judgment that determines what those patterns actually mean. The AI does not know which participant moments matter most for this decision. It does not weigh disconfirming evidence. It does not distinguish between a participant expressing a strong opinion and a participant agreeing passively with a prompt. What it produces is a useful starting point that still requires human nuance to add the context, interpretation, and perspective the compression removed. The failure mode is not using AI summaries. It is treating them as finished findings rather than as the first draft they are.

The most common version of this I heard in interviews was the AI smoothing away contradiction. A study has eight participants. Six say something positive about a feature. Two describe serious confusion. The AI summary says, "Participants generally responded positively, with some noting minor confusion." That framing buries the signal. The two confused participants might represent a real usability problem that affects a segment of users. The AI has no way to make that judgment. It just counts and averages.

How to address this: treat every AI-generated summary as a first draft, not a finding. Use it as a starting point to identify which transcripts to read in full, not as a substitute for reading them. When synthesizing, build evidence chains: every claim in the readout must trace back to specific participant moments, with quotes and context. If a claim cannot be traced, it does not go in the readout.

A practical rule I recommend: never let an AI summary be the last step before a readout. Always put a human synthesis step between the AI output and the stakeholder-facing document. The AI summary tells you what to look at. The researcher decides what it means. Those are different activities, and collapsing them into one step is where overclaiming happens. Some teams I interviewed described the same principle. One research lead put it this way: "AI finds the needles; I decide which ones matter and why." That division of labor works. Removing either side breaks it.

The third failure mode is the absence of an evidence trail. When a researcher conducts a traditional study, the evidence trail is built into the process: session recordings, timestamped notes, coded transcripts, and synthesis documents that link claims to data. When AI tools handle moderation and summarization, that trail can become thin or nonexistent if the researcher does not actively maintain it. This is improving. Tools like Outset now link summary claims back to specific participant quotes with timestamps, making it possible to trace a finding to the exact moment in an interview that produced it. That is a meaningful step. But the feature only works if the researcher actually uses it and if the organizational norm requires it.

How to address this: build the evidence trail deliberately. Store raw transcripts alongside AI summaries. When a finding goes into a readout, attach the specific participant moments that support it. Create a standard in which every directional claim has at least two traceable participant references. Several of the research teams I interviewed had implemented what they called "receipt requirements": no finding goes into a stakeholder-facing document without receipts. The discipline is simple. The effect on credibility is significant.

The fourth failure mode is moderation going off-script in unexpected ways. AI moderators generally follow instructions well, but they can produce follow-up questions that are outside the study scope, oddly phrased, or contextually inappropriate. This is less common with well-designed prompts, but it still happens, particularly when participants give unusual or ambiguous responses that the AI does not know how to handle. A human moderator would recognize the moment and make a judgment call. The AI guesses—and sometimes guesses badly.

How to address this: set clear boundaries in the moderation instructions about what topics the AI should and should not explore. Include explicit fallback instructions for ambiguous situations (for example, "If the participant gives an unclear response, ask them to say more; do not rephrase or interpret"). And again, the first-three-participants audit catches most of these problems before they propagate across the full sample.

The fifth failure mode—and this one came up in nearly every interview that touched on AI-assisted recruitment—is participant quality. AI tools make it easier to recruit at scale, but they also make it easier for low-quality or fraudulent participants to enter the sample: bot respondents, participants who speed through for incentives, participants who use AI to generate their own responses. The tools that make research faster also make fraud easier. This is an arms race, and researchers need to be on the right side of it.

The exposure level depends heavily on where your participants come from. If you are recruiting from your own user base—from people who actually use your product and can be verified against real accounts—the fraud risk is significantly lower. If you are recruiting from third-party panels, the risk goes up substantially. Panel providers vary widely in their quality controls, and the economics of panel incentives create a strong motivation for fraudulent participation.

How to address this: assume some percentage of your sample is compromised and build your workflow to catch it before it reaches synthesis. Review transcripts early, not just summaries. Pay attention to responses that are too clean, too fast, or too perfectly structured. Have a clear threshold for disqualifying a participant and a replacement protocol so the study does not stall when someone is removed. Some platforms, like Outset, are building fraud detection directly into the moderation layer, flagging low-quality or suspicious responses before they reach synthesis. That capability is improving and worth evaluating when choosing a tool. But platform-level detection does not replace researcher judgment. Speed of recruitment is not the same as quality of recruitment, and no tool fully solves both problems at once.

The Real Value, Despite the Limitations

Having been honest about the failure modes, here is the other side. The value these tools provide is structural. And dismissing AI-assisted research because of the limitations described above is like refusing to use email because spam exists. The limitations are real. They require mitigation. And the net effect, when the tools are used with appropriate judgment and quality systems, is a fundamental expansion of what research can contribute to product decisions.

Coverage increases. A researcher who could run two studies per month can, with AI-assisted moderation and synthesis support, contribute evidence to five or six decisions in the same period. The gap between decision velocity and evidence velocity shrinks. Researcher time reallocates to higher-value work. When AI handles moderation, transcription, and initial summarization, the researcher's time shifts to question design, quality review, interpretation, and decision communication. These are the parts where researcher skill matters most and speed becomes a choice, not a constraint. A question that needs depth can still get a two-week study with researcher-moderated sessions. A question that needs a fast directional answer can get a two-day study with AI-moderated sessions. The researcher chooses the mode based on the question, not based on the logistics.

The real comparison is not AI-assisted research versus a perfect traditional study. The real comparison is AI-assisted research versus what the team would do without it: guess, rely on internal opinion, misread analytics, or paste transcripts into ChatGPT and call it insight. Against that baseline, a well-designed AI-assisted study with real participants, quality gates, and evidence trails is dramatically better. The standard is not perfection. The standard is better than the alternative that will actually happen if research does not show up.

Where This Goes Next

This chapter mapped the capabilities and limitations of AI-moderated research as honestly as possible. The tools are real, the value is real, the limitations are real, and the judgment required to navigate between them is the core of the researcher's job in this new environment.

If there is one thing to take from this chapter, it is this: AI tools do not make research easier. They make research faster. Those are different things. Faster with judgment produces better outcomes. Faster without judgment produces confident garbage at scale. The systems and quality gates described in this chapter, and expanded in later ones, are what keep the work on the right side of that line.

The capabilities and limitations described here lead to a natural conclusion: AI-assisted research is not one thing. It is a set of tools that can be deployed at different levels of speed, scope, and researcher involvement depending on what the question requires. Chapter 4 introduces the engine that makes this operational: three modes, a routing logic for matching questions to the right mode, and the decision contract that locks scope before any study begins. The system starts there.

CHAPTER 4

Building the AI-Powered Research Engine

By this point in the book, the structural argument is established. Product teams generate decisions faster than research can inform them. The traditional research operating model does not match the tempo. AI tools provide real value but require quality systems and researcher judgment to be credible. The question that remains is operational: when a research request lands on your desk, what do you do with it?

That may sound like a simple question. It is not. The majority of research dysfunction I encountered in my interviews did not come from bad methods or bad tools. It came from misclassification. Questions that needed deep exploration got crammed into fast studies. Questions that needed a quick, directional answer got scoped into multiweek projects. Questions that were not research questions at all consumed researcher time because nobody had a system for saying, "This does not belong in our pipeline." The routing problem—deciding what kind of treatment a question deserves—turned out to be the most consequential decision in the entire research workflow. Get it right, and everything downstream works. Get it wrong, and no amount of methodological skill saves you.

This chapter introduces the taxonomy for routing research requests: three categories that cover the range of questions a research function encounters, with clear criteria for deciding which category a question belongs in. It also introduces the mechanism that locks scope once a question is routed: the decision contract. And it draws on a concept from Carl Pearson, whose work on Minimum Viable Rigor provides the intellectual backbone for matching rigor to risk in a way that is practical, not academic.[10]

Routing Questions to the Right Mode

Most research functions default to one speed. Everything gets treated the same way, with the same process, the same timeline expectations, and the same level of rigor. That default is where most research dysfunction starts. The routing framework in this book has three categories because three fundamentally different types of questions show up in a research pipeline, and treating them the same way produces the wrong outcome for all of them.

The two fast modes are defined by a structural constraint that reflects current tooling capabilities: they work for questions that can be answered through AI-moderated sessions, where participants respond to prompts, stimuli, or artifacts. Concept testing, comprehension checks, preference comparisons, objection mining, terminology validation—these fit naturally today. Methods that

10 Carl Pearson writes about Minimum Viable Rigor and related frameworks on his blog at https://www.carljpearson.com/.

require observing real-time behavior with a live interface, sustained engagement over days or weeks, or physical context are at the boundary of what current tools support, and that boundary is shifting. What will not shift is the routing logic: every question needs to be matched to the right speed and depth based on risk, ambiguity, and expiry. As the tooling expands, more methods will become available within the fast modes. The most common routing errors will still come from forcing a deep research question into a fast mode, regardless of what tools exist.

The three categories are micro research, sprint research, and deep research.

Micro research is for narrow, specific, near-term product questions where the decision is defined, the domain is understood, the ambiguity is low, and the question can be answered with real participants in twenty-four to seventy-two hours. The output is a directional conclusion with explicit boundaries on what it can and cannot support. Micro research is the fastest mode and the most constrained. It works precisely because the scope is tight. A litmus test: if the insight expires within a week, and the question is specific enough that you could explain it to a stranger in one sentence, it is probably a candidate for micro research.

The kinds of questions that fit micro research have a recognizable shape. Comprehension: do users understand what this screen is telling them? Friction: where do they get stuck in a three-step flow? Clarity: is this pricing explanation confusing? Preference: given two concrete options with clear trade-offs, which do users choose and why? Terminology: does this label land or backfire? Objection mining: what is the first reason someone would not do this? These are specific, concrete, and testable against real artifacts. If you can show someone a thing and get a useful answer in five minutes, it probably fits micro research. If the question requires extended conversation, contextual observation, or iterative exploration, it does not.

Sprint research is for questions with moderate ambiguity and moderate stakes that benefit from broader exploration over one to two weeks. The question may involve multiple related subquestions. The domain is generally understood, but the answer is not obvious. The researcher stays closely involved throughout, conducting follow-up sessions with participants, digging into emergent themes, and synthesizing across multiple inputs. Sprint research is AI-assisted but researcher-led. It is faster than traditional research because AI tools compress the execution, but it still requires sustained judgment across the arc of the study.

Deep research is for questions where the ambiguity is high, the problem space is not well understood, and the team members need to learn things they do not know they do not know. It is also the home for methods that require what AI-moderated sessions cannot currently provide: observing real-time behavior with a live interface, sustained engagement over days or weeks, or physical context. This is deep research: sitting with users for extended periods, observing behavior in context, building mental models, mapping workflows, and understanding motivation and identity. Deep research cannot be rushed. It requires the slow, human, often uncomfortable work of building understanding from scratch. If the question starts with "What do users actually need?" or "Why do people behave this way?," it is almost certainly deep research.

It is worth stating explicitly that the two fast modes exist because tooling such as Outset exists today. The taxonomy described in this chapter is not an abstract organizational framework that can be implemented with any set of tools. Micro research requires

AI-moderated data collection to hit its twenty-four- to seventy-two-hour timeline. Sprint research requires AI-moderated data collection to achieve the breadth of its first phase within days rather than weeks. Without the tooling layer, the taxonomy collapses to two categories: traditional research and deep research. That boundary will shift as tools improve. What will not shift is the routing logic: every question needs to be matched to the right speed and depth based on risk, ambiguity, and expiry. The specific tools that make each mode possible will evolve. The discipline of matching questions to the right treatment will not.

These three modes are not a menu. They are an engine. I call it the Research Engine, and it is the operational model this book is built around. Deep research builds the Frame: the foundational understanding of who the users are, how they think, what they need, and what the problem space actually looks like. Sprint and micro research operate within that Frame, answering specific questions with speed and credibility because the Frame tells you which questions are worth asking. And fast research signals when the Frame needs rebuilding, because the surprises, contradictions, and unexpected patterns that surface during micro and sprint studies are how you know the Frame is outdated or was never right to begin with. Deep research rebuilds. The cycle continues. That is the engine.

Most UXR organizations only run part of the engine. They operate within a Frame that was built years ago, or that was never built at all, and they wonder why their fast research keeps producing findings that surprise them or contradict each other. The answer is not better micro studies. The answer is that nobody maintained the Frame those studies operate inside—which means every finding was technically correct and potentially useless.

Chapter 5 covers what a maintained Frame requires and how to build one.

The rest of this book is about building and running the engine: how to route questions to the right mode, how to run each mode with credibility, how to protect the deep research work that keeps the Frame current, and how to scale the whole system without losing quality.

A note on time frames. The durations described above—twenty-four to seventy-two hours for micro, one to two weeks for sprint—are what these modes converge toward after you have run a few cycles, built your templates, established your recruitment channels, and trained your stakeholders on what to expect. They are not promises for day one. Your first micro study will not take seventy-two hours. It might take a week and a half. Your first sprint will not take two weeks. It might take four. That is fine. Your actual timelines will depend on your organization's UX maturity, your team's capacity, your recruitment access, your stakeholder norms, and how much governance infrastructure you have in place. The time frames tighten with practice and experimentation. Treat them as the goal the system is optimizing toward, not the prerequisite for starting.

The Research Engine at Different Stages of UXR Maturity

The operating model described in this chapter is the destination. It is not the starting point. A solo researcher at a Series B startup and a research director managing a fifteen-person function at a large tech company are both reading this book, and both can use the

Research Engine—but they are not starting from the same place, and they should not try to build the same thing on day one.

The engine has four components: the routing logic, the Frame, the three operating modes, and the governance structure. They are designed to work together. But they can also be adopted incrementally, in order of organizational urgency, without waiting for the full system to be in place before anything changes.

The sequence that works in practice is this.

Start with routing and the decision contract. These two things require no infrastructure, no additional headcount, and no organizational buy-in beyond the researcher's own practice. Routing changes how you handle the next incoming request. The decision contract changes what gets agreed before any study begins. A researcher who routes well and uses decision contracts will produce more useful work and encounter less scope creep within weeks of starting, regardless of whether anything else in the organization changes. This is the minimum viable version of the Research Engine, and it is available to anyone reading this book tomorrow.

Add the Frame when the team has capacity to own it. The Frame does not require a formal program from day one. It starts as an honest assessment: what does the organization currently believe about its users, how confident is the team in each belief, when was each part last updated, and who is accountable for knowing whether it still holds. That assessment can happen in a single working session. It will be uncomfortable. It will reveal gaps that nobody has named before. That discomfort is the point. A research team that has done this assessment, even informally, is operating from a fundamentally different foundation than one that has not. The structured document, the coverage matrix, the belief map—those come later. The assessment comes first.

Add governance when the volume justifies it. One researcher running all the studies does not need an organizational governance policy. They need personal quality habits: transcript auditing, evidence chains, disconfirming evidence checks, and expiry dates. Those habits are governance for a team of one. When the volume grows, when other people start running studies, and when the organization needs shared quality standards rather than personal discipline, the governance infrastructure from Chapter 9 becomes necessary—not before.

Protect deep research from the beginning, not eventually. This is the one exception to the incremental logic. Deep research does not get easier to protect as the organization matures. It gets harder, because the demand for fast work grows faster than the capacity to run it. A researcher who waits until the organization is mature enough to appreciate deep research will wait indefinitely. The Frame-level question log, the calendar blocking, the named sponsor—these are practices that any researcher can start immediately, regardless of organizational maturity. The protection mechanisms do not require permission. They require discipline.

The table below describes what the Research Engine looks like at three broad levels of organizational maturity. These are not rigid stages; they are reference points. Most organizations are somewhere between levels, strong on some components and weak on others. Use the table to identify which components are already in place and which represent the highest-leverage next investment.

Low UXR maturity: Research as a single pipeline

The research function operates on one speed. All requests go into the same process. Quality depends on individual researcher practice rather than shared standards. The Frame exists implicitly but has not been named or assessed. Deep research happens when someone champions it, not on a cadence.

What to build first: routing and the decision contract. Start saying, "This is a micro question; here is what I can deliver and when," instead of treating every request the same way. Write a decision contract for every study. Even if nobody reviews it, writing it forces the scoping discipline that prevents the most common quality failures.

What to defer: formal governance, the coverage matrix, and the belief map. These require organizational infrastructure that low-maturity teams do not yet have.

Medium UXR maturity: Research with some structure

The research function has a routing instinct, even if it is not formalized. The decision contract is standard practice. The Frame has been assessed at least once and has informal ownership. Some governance exists as a team habit. Deep research happens but is not structurally protected.

What to build next: make the Frame explicit. Document current organizational beliefs by segment and domain. Assess coverage, freshness, and confidence. Establish clear ownership. This transforms the Frame from informal shared understanding into a maintained asset. Also build the Frame-level question log if it does not exist. Track the signals that fast research generates about where the Frame is getting stale.

What to defer: conditional self-serve and the full governance policy. These require a volume of research that medium-maturity teams typically are not yet running at scale.

High UXR maturity: The full Research Engine

The routing logic is internalized across the team. The decision contract is organizational policy. The Frame is explicitly maintained with defined ownership and a regular synthesis cadence. Governance exists as shared standards that apply to anyone running research. Deep research is structurally protected through calendar blocking, named sponsors, and a defined budget. All three modes are running simultaneously.

What to focus on at this level: Frame quality across the organization. In large organizations with multiple product areas and overlapping user populations, the federated governance question we'll discuss in Chapter 5 becomes central. Who has visibility across the Frame? Where are different product areas operating on contradictory beliefs about the same users? How does the Frame steward coordinate across embedded research teams? These are the questions that high-maturity organizations need to answer—and that lower-maturity organizations can defer.

One practical note on sequencing. The components that deliver the most immediate value—routing and the decision contract—are also the easiest to start. The components that deliver the most long-term value—the Frame and deep research protection—are the hardest to sustain. The temptation is to skip the easy components because they seem too simple and focus on the ambitious ones because they feel more strategic. This is backward. The easy components build the credibility and the organizational habits that make the ambitious ones possible. Start where the friction is highest and the value is most immediate. Everything else follows from there.

When the Problem Is Not UXR Maturity but Legitimacy

The maturity progression above assumes that research is already a recognized function with some organizational standing. Questions arrive. Stakeholders engage. The routing logic has somewhere to operate.

Not every researcher is in that environment. Some are in organizations where research is new, where stakeholders do not yet know what to ask for, and where the first challenge is not routing but being in the room when the questions get asked at all. In those environments, the Research Engine is still the right destination. But the entry point is different.

The legitimacy problem shows up in a specific way. A researcher reads the routing logic and thinks, "This makes sense, but nobody brings me questions I can work with." Requests arrive as solutions looking for validation or as vague mandates with no decision attached, or they do not arrive at all because the team has learned to make decisions without research and has not been given a reason to stop. The framework described in this chapter assumes a working relationship between researcher and stakeholder. Building that relationship is the prerequisite, and in low-legitimacy environments it is the actual first problem to solve.

The practical approach is to start with visibility rather than routing. Attend planning meetings without being asked. Listen for decisions being made without evidence. When one happens, note it privately. Over a few weeks, build a small log of moments where user evidence would have changed the outcome, or might have. That log is not for public use yet. It is for building the internal case that research belongs in the conversation.

The second step is to make one routing decision visible and explain it. When a request arrives—even an imperfect one—respond not just with a timeline but also with the reasoning. "This is a comprehension question with low stakes and a tight deadline, so I am going to run a focused study with eight participants and have directional findings by Thursday. It will not tell us whether users will adopt the feature, but it will tell us whether they understand the flow." Those statements do two things: they deliver something useful, and they teach the stakeholder what research can and cannot do, without a lecture. Do that consistently, and the nature of the requests will start to change.

The third step is to protect one piece of proactive work—even a small one. A researcher who only responds to requests will always be reactive. Finding 20 percent of their time, however difficult, to investigate something the organization has not asked about yet, and then delivering that investigation in a form that changes how the team thinks about a problem, is the fastest path from reactive to legitimate. It does not have to be deep research. It can be a micro study on something the team has been guessing about for months. The point is not the method. It is demonstrating that research surfaces things the organization did not know it needed to know.

The routing logic, the decision contract, the Frame, the governance infrastructure—all of it becomes possible once legitimacy is established, not before. In low-legitimacy environments, the Research Engine is the map. Building the trust to use it is the terrain.

Matching Rigor to Risk: The Routing Logic

The question of which category a request belongs in is ultimately a question about risk and rigor. How much is at stake if the team gets this wrong? How much ambiguity exists in the question? How much rigor does the answer need to be credible enough to act on?

Carl Pearson's concept of Minimum Viable Rigor (MVR) provides a useful framework here. Carl argues that rigor in applied research is not about academic purity. It is about decision quality. The goal is not perfect truth. The goal is to align closely enough with truth that the team's actions achieve their intended outcomes at an acceptable rate, given the risk tolerance of the decision being made. MVR is the threshold below which an insight cannot reliably support the decision it is meant to inform. Above the line, the evidence is good enough to act on. Below it, you are making decisions based on something that looks like evidence but will not hold up.

Carl's formula is elegantly simple: MVR = Rigor of Insight – Decision Risk. When the stakes go up, the rigor requirements go up. When the stakes are low, you can move faster with less rigor and still produce something credible. This is not a license to be sloppy. It is a license to be proportionate.

What I did in developing the operating model for this book, and what I tested with practitioners in my interviews, was to translate Carl's framework into a practical routing logic. The key insight is that the same dimensions that determine MVR also determine which mode a question belongs in. Risk, ambiguity, and expiry are not just abstract properties of a question. They are routing criteria.

Here is how they work in practice.

Risk refers to what happens if the team gets the answer wrong. A copy change on a low-traffic settings page carries low risk. A redesign of a core revenue flow carries high risk. A new product entering a regulated market carries very high risk. The risk level determines how much rigor the answer needs. Low-risk decisions can tolerate directional evidence with explicit limitations. High-risk decisions need converging evidence from multiple sources, larger samples, and more careful analysis.

Ambiguity refers to how well the question is defined and how well the domain is understood. A question like "Do users understand what this button does?" is low ambiguity. A question like "What do small-business owners actually need from a financial dashboard?" is high ambiguity. Low-ambiguity questions can be answered quickly because the research design is straightforward. High-ambiguity questions require exploration, follow-up, and the kind of iterative understanding that cannot be compressed into a day or two.

Expiry refers to how long the answer will be relevant. Some questions expire in a week because the feature is moving into development next Tuesday regardless. Some questions are relevant for a quarter. Some inform strategic direction that will shape the road map for a year. Short-expiry questions need fast answers. Long-expiry questions justify deeper investment because the findings will be used longer and influence more decisions.

The routing logic maps these three dimensions to the three categories.

Low risk, low ambiguity, short expiry: micro research. The question is clear, the stakes are modest, and the answer is needed fast—a twenty-four- to seventy-two-hour AI-moderated study with real participants, a tight prompt set, and a directional readout with explicit boundaries.

Moderate risk, moderate ambiguity, medium expiry: sprint research. The question is understood but not obvious. The stakes are meaningful. The answer will inform a decision that matters for the next few weeks or months—a one- to two-week, researcher-led, AI-assisted study with follow-up sessions, broader sampling, and more careful synthesis.

High risk, high ambiguity, long expiry: deep research. The team does not understand the problem space well enough to ask a specific question. The stakes are significant. The findings will shape strategy. This needs research with extended user engagement, contextual observation, and sustained analysis. Once deep research has mapped the problem space, AI-moderated micro and sprint studies can systematically address the subquestions it surfaces. Deep research defines the territory. Fast research explores within it. That division of labor is the engine working as designed.

Some requests are not research questions at all. They are measurement questions that belong with analytics or data science. Recognizing them quickly and redirecting them is part of the routing discipline.

This is not an algorithm. It is a thinking tool. The dimensions do not always align neatly. A question might have moderate ambiguity but high risk, which pushes it toward sprint research rather than micro, even though the expiry is short. A question might have low risk but high ambiguity, which means deep research is the right mode even though the stakes do not seem to justify the investment. The researcher's judgment is what navigates these tensions. The taxonomy gives that judgment a structure to work within.

One important nuance: the dimensions can change mid-study. A question that looked like micro research at intake can reveal enough ambiguity during the first three participant reviews that it needs to be escalated to sprint research. A sprint study can surface something so unexpected that it triggers a deep research project. The taxonomy is not a one-time classification. It is a continuous assessment. The decision contract helps here too, because it establishes what the study will and will not conclude. If the study reveals that the question is bigger than expected, the contract gives the researcher a clear basis for saying, "This is beyond what we scoped, and here is why we need to escalate."

Sizing Risk in Practice

The risk dimension deserves more attention because it is where most routing disagreements happen. Researchers and stakeholders often have different intuitions about how risky a decision is, and those different intuitions lead to different expectations about how much rigor is needed.

Building on Carl's MVR framework, I found it useful to break risk into several contributing factors rather than treating it as a single judgment. These are impact on users (how many people are affected and how significantly), business exposure (revenue, cost, or strategic consequences of being wrong), reversibility (how easy it is to undo the decision if it turns out to be wrong), scope of exposure (how widely the change will be visible), and regulatory or compliance sensitivity (whether the decision touches legally sensitive territory).

None of these factors alone determines the risk level. A change that affects many users but is easily reversible is lower risk than a change that affects fewer users but is hard to undo. A decision with significant revenue implications but low compliance sensitivity is different from one with modest revenue impact but high regulatory exposure. The factors interact, and the researcher's job is to weigh them in context.

The practical benefit of breaking risk into factors is that it makes the conversation with stakeholders concrete. Instead of debating whether something is "high risk" or "low risk" in the abstract, you can walk through the factors together. For example: "This change affects our core checkout flow, so business exposure is significant. But it is behind a feature flag, so reversibility is high. And it does not touch anything regulated, so compliance sensitivity is low." That conversation produces a shared understanding of risk that neither the researcher nor the stakeholder could reach alone.

A research lead at a consumer fintech company described introducing this approach to structured risk conversations into her team's intake process. Before, risk was a gut feel that produced constant disagreement. Stakeholders would say, "This is low risk; just do a quick study," and the researcher would say, "Actually, this touches our payments flow, so it is high risk." Neither could convince the other because they were talking about different aspects of risk. After introducing the factor-based conversation, the disagreements decreased because both sides were looking at the same dimensions. "We still disagree sometimes," she said, "but now we disagree about specific things, not vague feelings. That is much more productive."

What matters here is the principle: risk is multidimensional, and routing decisions improve when you break risk into its components rather than treating it as a single intuitive judgment.

The Decision Contract: Locking Scope Before You Start

Routing a question to the right category is the first step. The second step is making sure the scope stays locked once the study begins. This is the job of the decision contract.

A decision contract is a short document, typically one page, that is agreed upon between the researcher and the stakeholder before any research begins. It captures five things.

First, the decision. What specific product decision will this research inform?

Second, the decision owner. Who will act on the results? This is a named individual, not a team. If nobody is willing to put their name on it, the decision is not real, and the study should not run.

Third, the outcomes. What will this research inform, and how will the team use it? For some studies, this is concrete: if the study shows X, we will do A; if it shows Y, we will do B. For others, the outcome is directional: the team will use the findings to inform a decision that has not been fully shaped yet, or to narrow a set of options, or to build confidence in a direction before committing resources. Both are legitimate. What is not legitimate is running a study with no connection to any decision or direction at all. The decision contract does not require that every study map to a binary action. It requires that someone can articulate why this research matters and what it will be used for. If the answer is "we just want to know," that is usually a sign that the question has not been scoped well enough, or that the real purpose is political cover rather than evidence.

Fourth, the boundaries. What will this study not conclude? This is the section most researchers skip and most stakeholders ignore, and it is the most important one. A micro research study with twelve participants cannot conclude anything about population prevalence. A sprint study focused on one market cannot generalize to other markets. A comprehension test cannot tell you whether users will adopt the feature. Stating boundaries explicitly in advance prevents findings from being stretched beyond what they can support.

Fifth, the expiry date. When do these findings stop being valid? Fast research produces directional signal with a shelf life. A micro research finding about user comprehension of a specific screen is valid until that screen changes. A sprint research finding about workflow friction is valid until the workflow is redesigned. Expiry dates prevent old findings from being cited as current truth months or years after the context has changed.

The decision contract is protection. It protects the researcher from scope creep. It protects the stakeholder from overclaiming. It protects the organization from treating directional findings as a permanent strategy. And it takes fifteen minutes to write.

Common Misclassifications and How They Fail

With the taxonomy and routing logic established, it is worth spending time on the ways it goes wrong. Because even with a clear framework, the organizational pressure to classify everything as fast research is strong, and the consequences of misclassification are real.

The most common misclassification, and the most damaging, is cramming a deep research question into micro research. This happens when a team has a genuinely open, exploratory question but frames it as a narrow, fast study because the timeline demands speed. "What do users need from our new dashboard?" becomes "Do users understand these three widgets?" The micro study runs. The three widgets test fine. The team concludes the dashboard concept is validated. It is not. The team tested comprehension of three components. It did not test whether the dashboard solves a real problem, whether users would actually use it, or whether the information hierarchy matches how users think about the domain. The micro study answered the micro question correctly. The team interpreted it as an answer to the macro question. That is a misclassification failure.

The inverse misclassification also happens: treating a micro question as if it needs a full study. This is less dangerous but more wasteful. A PM asks whether users understand a new label on a button. The researcher scopes a two-week study with twelve participants, a full discussion guide, and a synthesis deck. The question could have been answered in a day with five participants and a concrete task. The label works or it does not. By the time the readout lands, the team has already shipped with the old label because it could not wait. The research was rigorous and irrelevant.

A third misclassification is routing instrumentation questions through the research pipeline. A stakeholder asks the researcher to "look into" why a metric dropped. The researcher pulls data, builds charts, and presents an analysis. This may be useful work, but it is not a research study, and it consumed time that could have been spent on questions that required actual user contact. In mature organizations, this gets caught by intake triage. In less mature organizations, researchers become data analysts by default because nobody else is doing it and saying no feels political.

A fourth, subtler misclassification is treating a high-risk question as medium-risk because the timeline is tight. This is where the MVR framework is most valuable. The risk of a decision does not decrease because the deadline is soon. A question about a core revenue flow is high-risk whether the team has four weeks or four days. If the timeline does not allow sufficient rigor for the risk level, the right answer is not to do fast research with insufficient rigor. The right answer is to flag the gap, describe what the research can and cannot conclude given the constraints, and let the decision owner choose whether to proceed with limited evidence or adjust the timeline. This is uncomfortable. It is also the researcher's job.

Carl Pearson's MVR framework gives researchers the language and the logic to have this conversation without it becoming personal. "The risk level of this decision requires a certain level of rigor. The timeline allows this much rigor. There is a gap. Here is what we can do within the timeline, and here are the explicit limitations on what those findings can support." That is a professional assessment, not a refusal. It puts the choice where it belongs: with the decision owner, who has the context to weigh the risk of acting on limited evidence against the risk of waiting.

How Routing Actually Happens

It would be nice to say that routing happens through a clean intake form, a structured risk assessment, and a collaborative decision contract conversation. And in some organizations, it does. But in my interviews, the reality was messier and more varied than any single process could capture.

Some teams used formal intake systems. A request came in through a form, got triaged by the research lead, and was routed to the appropriate mode with a written rationale. These teams tended to be larger, with dedicated research operations support. The advantage was consistency: every request got the same evaluation. The disadvantage was speed: the intake process itself added a day or two to the timeline, which, for micro research questions, meant the answer arrived after the decision had already been made.

Some teams routed through conversation. The researcher sat in planning meetings, heard questions as they surfaced, and made real-time judgment calls about what to investigate and how. These teams tended to be smaller, with researchers embedded closely in product pods. The advantage was responsiveness: routing happened instantly. The disadvantage was inconsistency: the routing was only as good as the individual researcher's judgment in the moment, and there was no audit trail for why a question was treated one way rather than another.

Some teams used a hybrid approach. Quick questions got routed conversationally. Anything that would take more than a day or two went through a lightweight intake process. The decision contract was required for sprint research and above but optional for micro research (though smart researchers wrote one anyway, because the discipline of naming the decision and the boundaries was worth the five minutes it took).

The right approach depends on the team, the culture, and the volume of requests. What matters more than the specific mechanism is that routing happens deliberately rather than by default. The failure mode is not choosing the wrong process for routing. The failure mode is not routing at all, which means every request gets treated the same way, usually as whatever the researcher happens to have capacity for, with no explicit assessment of risk, rigor, or fit.

Where This Goes Next

This chapter gave you the taxonomy, the routing logic, and the decision contract. Together, they form the operating system for deciding what to do when a research request arrives: micro research, sprint research, deep research, or not research at all, matched to risk, ambiguity, and expiry and locked with a contract that prevents scope creep and misuse.

Chapter 5 covers the Frame in full: what it is, how to assess its current state, and what maintaining it actually requires. Chapter 6 covers sprint research: how to scope it, design studies within it, run AI-assisted sessions with researcher-led follow-ups, and synthesize across a broader evidence base in one to two weeks. Chapter 7 covers micro research: the twenty-four- to seventy-two-hour operating mode for narrow questions, including prompt design, modality selection, quality gates, and directional readouts with expiry dates.

But the taxonomy comes first because, without it, the modes are just labels. With it, they are part of a system that matches questions to the right treatment and protects each mode from being misused. The routing decision is the highest-leverage moment in the research workflow. Everything that follows depends on getting it right.

And one final thought. The taxonomy also protects the researcher. Without a routing framework, every request feels like a personal negotiation. Should this be fast? Should this be deep? Am I being too slow? Am I cutting corners? With a taxonomy backed by risk assessment and locked by a decision contract, the researcher is not making a personal judgment call in isolation. They are applying a shared system that both they and their stakeholders agreed to. That changes the dynamic from "the researcher is being difficult" to "the system is telling us this question needs more than two days." The system absorbs the organizational pressure so the researcher does not have to.

CHAPTER 5

The Frame

Chapter 4 introduced the Research Engine and the concept at its center: the Frame. Deep research builds the Frame. Sprint and micro research operate within it. Fast research signals when the Frame needs rebuilding. That is the operational logic of the system.

But the Frame itself deserves more than a paragraph. It is the most consequential concept in this book, and treating it as a given understates both what it is and how difficult it is to maintain. Organizational leaders who read Chapter 4 and concluded that they need to run more deep research have missed half the argument. The question is not just how often you build the Frame. It is who owns it, how you assess its current state, what it means to maintain it over time, and what happens when it degrades—which it always does, and at a pace that almost nobody notices until the damage is already done.

This chapter is the full treatment.

Why the Frame Matters

The Frame matters because it determines whether fast research is useful or just fast. Micro and sprint research answer questions. The Frame determines whether those questions are the right ones. An organization without a functioning Frame can run studies continuously, produce technically sound findings, and still make systematically wrong product decisions because every study is operating within untested assumptions. The research is not failing. The foundation it is sitting on is.

But there is a deeper reason. The standard research operating model treats knowledge as something that arrives in events. A study runs. A deck gets presented. Findings enter the organization. And then time passes, conditions change, and the knowledge stops being true—but nobody notices because the event is over and the calendar has moved on. The repository fills up. The archive grows. And the organization keeps operating on understanding that is quietly expiring, because nothing in the model tracks its decay.

The Frame is the countermodel. It treats user knowledge not as a series of deliveries but as a continuous state the organization maintains. Research output stops being a deck and becomes a delta: here is what changed in our understanding, here is what got confirmed, here is what we thought was true and no longer is. That shift—from knowledge as event to knowledge as state—is the architectural change the Frame represents. Everything else in this chapter follows from it.

That is not a methods problem. It is not a tooling problem. It is a structural problem, and it is the most expensive kind because it is invisible. Everything looks productive. Studies run. Decks get delivered. Findings get presented. The dysfunction only surfaces when something ships and fails in ways the research did not predict, by which point the decisions have already been made and the road map has already moved on.

What the Frame Is Not

The Frame is easy to confuse with things that already exist and already have names. That confusion is worth clearing up before defining the concept, because the most common response when the Frame is introduced is recognition followed by misidentification. "Oh yes, we have that. It's our repository." Or "That's our persona set." Or "We call that our journey map."

In almost every case, they don't have it.

A repository is an archive of what research has been done. It stores studies, findings, decks, and recordings. It tells you what happened. The Frame is about what the organization currently believes. Those are different things, and treating them as interchangeable is how organizations end up with extensive, well-organized research archives and no shared understanding of their users. The repository answers the question: what did we study? The Frame answers the question: what do we know right now?

A persona set is a representational artifact. It is a way of communicating user archetypes, usually created at a point in time and usually for a specific audience or planning purpose. Personas can express parts of the Frame. They are not the Frame. Most persona sets are partially inferential, built from a synthesis that reflects a specific research effort at a specific moment, and entirely static once created. The Frame is neither of those things.

A journey map is a process visualization. It is useful for documenting how users move through an experience, identifying friction points, and aligning teams on flow. It is a slice of the problem space. The Frame is the whole problem space. It is not just how users navigate a specific flow but also who they are, what motivates them, what their broader context looks like, and how the product fits into their lives.

A research strategy is a plan for what the team intends to study. The Frame is what the organization currently knows. A team can have a well-designed research strategy and an outdated Frame. In fact, this is common. The strategy documents what the team plans to do. The Frame reflects what the organization actually understands. When these diverge, the strategy produces technically sound work that somehow never quite moves the needle because it is optimizing inside a model of the user that no longer holds.

A Definition

The Frame is the organization's accumulated, actively maintained model of its users.

Not what it has studied, but what it currently believes, based on the best available evidence, about who its users are, what they are trying to do, what motivates them, what creates friction for them, how they make decisions, and where the product fits—or does not fit—into their lives.

Three words in that definition carry significant weight.

Accumulated means the Frame is built over time from multiple sources, not generated by a single study or owned by a single team. It incorporates qualitative findings, quantitative patterns, behavioral data, support signals, market research, and product intuition when that intuition is grounded in real observation. UXR does not own the Frame exclusively; UXR stewards it. That distinction matters for governance and for organizational buy-in, both of which are addressed later in this chapter.

Actively maintained means someone is responsible for keeping it current. Not passively stored and not allowed to sit until a major initiative creates an opportunity to revisit it, but maintained on a cadence, with explicit attention to where it is getting stale, where coverage is thin, and where the organization is operating on assumptions that have not been tested in a meaningful amount of time. Most organizations have a version of the Frame that was built at some point and has been degrading ever since. The Frame that exists but is not maintained is not the Frame this chapter is describing.

Model means it is a structured representation with internal coherence, not a pile of findings. A model is something you can use to make predictions. If the Frame is accurate, it should help the organization anticipate how users will respond to something new, not just describe how they responded to something the team shipped two years ago. Most research functions produce descriptive knowledge: here is what users said, here is what users did. The Frame, when it is working, produces predictive capacity: given what we know about these users, here is what we should expect. That is a different kind of output, and it is the kind that justifies the ongoing investment that Frame maintenance requires.

The Four Properties

The Frame has coverage, freshness, confidence, and ownership—four properties that you can assess explicitly. Not with perfect precision, but with enough specificity to move from vague conversation about "Do we understand our users?" to a concrete discussion about which parts of that understanding are strong, which are degraded, and who is responsible for the gap.

Coverage is which parts of the user population and problem space the organization understands well versus which parts it is operating on assumption. Most organizations have deep coverage in a few areas, typically wherever the founding team had direct personal experience or wherever the last major research effort happened to land. The gaps rarely announce themselves. They quietly shape every decision made in their vicinity, and nobody notices until a product ships into an undercovered space and fails in a way that feels inexplicable. A coverage assessment asks: for each meaningful user segment and each significant part of the problem space, how confident are we that we understand it well?

Freshness is when each part of the Frame was last meaningfully updated. Freshness is not uniform across the Frame. Some parts of a user model age in months. Others are stable for years. A study on how users navigate a specific flow might still be accurate two years later if the flow has not changed significantly. A study on what motivates users to try a new product category might be stale in six months if the competitive landscape has shifted. Treating all past research as equally current or equally suspect is both analytically wrong and operationally expensive. A freshness assessment asks: for each part of the Frame, when was the underlying evidence generated, and is that evidence still likely to reflect current reality?

Confidence is the degree to which each belief in the Frame is based on strong direct evidence versus inference, extrapolation, or organizational folklore. Every organization has beliefs about its users that have achieved the status of common knowledge without anyone being quite sure where they came from. A researcher said it in a readout once. It then appeared in a strategy deck. A PM repeated it in a planning meeting. It got cited in the next planning meeting as though it were an established finding. Now it is true. Confidence assessment makes those beliefs visible and distinguishes between "we have strong, recent evidence for this" and "we believe this, but the basis is unclear." Both can inform decisions, but they carry different organizational risk.

Ownership is who is accountable for knowing whether the Frame still holds. Not who ran the last study on a given topic; rather, who is responsible, right now, for knowing whether a specific part of the Frame is current and accurate—and for calling for updated work when it is not. Ownership is the property that makes the other three actionable. Without it, coverage assessment produces a list of gaps that nobody has the mandate to address. Freshness assessment produces a list of stale beliefs that nobody has the authority to flag. Confidence assessment produces an uncomfortable conversation that dissipates when the meeting ends because nobody owns the follow-through.

Most organizations can, with effort, piece together answers on coverage and freshness imperfectly if they invest the time to look. Ownership is different; it is rarely defined in a clear, operational way. And that is the property that matters most. Without ownership, gaps remain unaddressed, stale beliefs persist, and weak assumptions continue to drive decisions. As a result, ownership is often addressed last—usually only after something has already gone visibly wrong.

Why Frames Degrade

The degradation of the Frame is almost never the result of a deliberate decision. Nobody holds a meeting and decides to stop maintaining the organization's model of its users. It happens incrementally, through a series of individually reasonable choices that collectively produce a belief system that no longer reflects reality.

Most organizations had a functional version of the Frame at some point. There was a foundational study, or a segmentation initiative, or a generative research effort that a previous leader championed—and that actually changed how the organization thought about its users. For a period of time, the organization operated from that understanding. Decisions were sharper. Research felt purposeful in a way it often does not, because there was something substantial to connect findings to.

Then the team got busy.

The Frame from that effort is still technically present. It lives in a deck somewhere or in the institutional memory of a few researchers who were there when it was built. But it is not current. The users changed. The product expanded into territory the original Frame never covered. The competitive landscape shifted. New user segments emerged that nobody mapped. But the organization is still operating within the old Frame implicitly, because nobody has had the bandwidth or the organizational permission to rebuild it.

The mechanism of degradation is specific. Research continues, but it runs within the old Frame rather than questioning it. Micro and sprint studies answer questions that assume the Frame is correct. The Frame-level assumptions—who the users are, what they want, what problems the product solves for them—are treated as settled rather than as claims that require ongoing verification. Over time, the gap between the Frame and reality widens. The fast research gets more precise and less useful. Findings that are technically correct do not produce the expected outcomes.

The symptoms are recognizable. The team keeps getting surprised by user behavior that should not happen given their understanding. Sprint studies surface the same confusion from different angles, suggesting the confusion is not about a specific screen but about a foundational misunderstanding between the product and its users. Features launch with strong comprehension test results and then fail to drive adoption. The numbers say users understand. The behavior says they do not care. That gap is almost always a Frame problem.

None of those symptoms are diagnosable through micro or sprint research. An organization can run 100 well-executed fast studies and never discover that the Frame is wrong, because fast studies test within the Frame. Discovering that the Frame is wrong requires stepping outside it, which is the work of deep research.

The Frame and the Research Engine

It is worth being precise about the relationship between the Frame and the three research modes introduced in Chapter 4, because this relationship is the operating logic of the whole system.

Deep research builds the Frame. It produces the foundational understanding of who the users are, what their context looks like, what motivates them, what problems they have, and where the product fits into their lives. When the Frame does not exist, or when it has degraded past the point of usefulness, deep research is how you rebuild it. This is why protecting deep research—which Chapter 10 addresses in detail—is not optional. It is the mechanism by which the Frame gets established and renewed.

Sprint and micro research operate within the Frame. They answer specific questions with speed and credibility because the Frame tells them which questions are worth asking. A sprint study on user comprehension of a pricing structure assumes that the pricing model itself is solving a real problem for real users. A micro study on navigation assumes that the thing users are navigating toward is something they want. Those assumptions are Frame-level claims. If they are wrong, the sprint and micro findings are technically correct and strategically useless.

Fast research signals when the Frame needs rebuilding. When micro and sprint studies keep producing contradictory findings, or when findings are consistently accurate but consistently fail to predict behavior, or when users keep responding to questions with confusion that goes beyond the scope of the question, those are signals that the Frame-level assumptions are no longer holding. Chapter 10 covers the specific practice of maintaining a Frame-level question log: a running record of the moments in fast research where something does not add up, which collectively build the case for when deep research needs to happen.

Deep research rebuilds. When the signals accumulate, or when the product moves into new territory, deep research updates the Frame. The new Frame enables better fast research. The cycle continues. When the cycle breaks—because the organization stops protecting deep research or because ownership of the Frame disappears—the fast work continues but gradually becomes less and less useful, and nobody can explain why.

What the Frame Looks Like in Practice

The Frame is a model, not a single artifact. In practice, it is expressed through four representations that serve different purposes and different audiences. Understanding what each one contains, and how they relate to each other, is what makes the concept operational rather than abstract.

The first is the structured document. This is the Frame itself. It is organized by user segment or problem domain, whichever division is most meaningful for the product. Each entry contains the current organizational belief about that segment or domain; the evidence base that supports it; a freshness indicator showing when the underlying evidence was generated; a confidence level distinguishing strong direct evidence from inference or folklore; and a named owner accountable for keeping that entry current. The structured document is maintained by the researcher or research team closest to that product area. It is not a deliverable. It is a working document that gets updated as understanding changes, with a record of what was believed before and what superseded it.

Where the structured document lives matters as much as what it contains. A Frame that sits in a research repository that product and engineering teams never open is functionally unavailable. The interaction layer—how teams actually reach the Frame in the flow of their work—is an operationalization problem that most research functions get wrong because they focus on building the Frame and assume people will find it.

The most effective interaction layers share a common design principle: the Frame should be findable where decisions are already being made, not in a separate system that requires a context switch.

A product team that plans in Notion should find Frame references in Notion. A team that builds specs in Confluence should find Frame links in Confluence templates. A team that uses a design system should find user understanding embedded in the component documentation. This is not about duplicating the Frame everywhere. It is about creating lightweight pointers from the places where decisions happen back to the relevant Frame entry.

Three mechanisms work in practice. The first is embedded links in planning templates. When a PM opens the product brief template, the first field after the problem statement should be a link to the relevant Frame entry. Not a copy of it. A link. The PM clicks it, reads the current organizational belief about the users this decision affects, and scopes the work accordingly. The template enforces the habit. The second is Frame coverage in research readouts. Every readout from a micro or sprint study should include a one-line Frame reference at the top: which Frame entry the study operates within, and whether the findings update it. This connects fast research output back to the Frame in real time rather than requiring a separate synthesis step. The third is a standing Frame review in planning cycles. Once per quarter, the Frame steward presents a ten-minute summary of what changed in the Frame since the last planning cycle, what gaps are most significant given the upcoming road map, and what research is planned to address them. This is not a readout. It is a reference briefing. It keeps the Frame visible without requiring everyone to maintain it themselves.

The underlying principle behind all three mechanisms is the same: research knowledge that lives in research-owned spaces gets used by researchers, and research knowledge that lives in product-owned spaces gets used in product decisions. A study update placed directly in the planning surface where the PM is working will be read. The same update filed in a research repository will not. That is not a small operational detail. It is what determines whether the knowledge reaches the decision or expires in an archive.

The goal is not to make everyone a Frame steward. It is to make the Frame impossible to ignore at the moments that matter. The interaction layer is the infrastructure that makes that possible.

The second mechanism is the coverage matrix. This is the governance tool. It maps user segments against problem domains and shows, at a glance, where the organization has strong coverage, where coverage is thin, and where it is operating on assumption. Each cell reflects the freshness and confidence of the underlying structured document entry. The coverage matrix does not contain beliefs. It contains an assessment of how well-founded those beliefs are. It is most useful at the organizational level, maintained by whoever owns Frame governance across product areas, and reviewed on a cadence to identify where discovery work is needed before a product decision exposes the gap.

In practice, the coverage matrix is most useful when rendered as a heatmap: user segments on one axis, problem domains or product features on the other, with each cell shaded according to evidence density. Hot cells are where the organization has strong, recent, confident evidence. Cold cells are where it is operating on assumption. The heatmap makes the gaps viscerally visible in a way a spreadsheet does not. A cold quadrant in a product area about to receive significant investment is a risk conversation that needs to happen before the investment is made, not after.

Building the heatmap requires two passes—one top-down and one bottom-up—and both are necessary.

The top-down pass comes first. Draw the grid before you look at any research. Map your user segments against your product verticals, features, or problem domains, whichever division is most meaningful for the product. This gives you the complete picture of what the organization should understand. The cells that matter are not just the ones where research happened. They are all of them, including the ones that will be empty. The top-down pass makes the cold areas visible by design rather than by accident.

The bottom-up pass populates the grid. Go through every study the team has run and place it on the grid. A cell that has been studied recently with a method that matches the question and has produced confident findings is hot. A cell that has only micro studies—which tested behavior but never explored motivation or mental models—is lukewarm. A cell where the last study was two years ago and the product has changed significantly since then is cooling. A cell where nothing has ever been done is cold. The bottom-up pass gives you an honest picture of where the evidence actually sits rather than where you think it should.

The Frame steward's job is then to read the heatmap against what the product is doing next. A cold cell in a stable, low-investment area is a manageable gap. A cold cell in an area where a significant bet is about to be made is a risk that needs to be named before it becomes expensive. The coverage matrix does not make those calls automatically. It makes them visible so the right conversation can happen at the right time.

The third mechanism is the belief map. This is the communication artifact. It translates the structured document into a form that is legible to leadership, product partners, and cross-functional stakeholders who need to understand what the organization currently believes about its users without reading the underlying documentation. The belief map organizes the most important organizational beliefs visually, with confidence indicators that make the distinction between strong evidence and inference visible. It is not a replacement for the structured document. It is a synthesized view of it, produced when stakeholder alignment or strategic planning requires a shared picture of current organizational understanding.

The Assumption Audit

The fourth mechanism is the assumption audit. This is the diagnostic tool. Where the coverage matrix shows where evidence exists and where it does not, the assumption audit shows something more dangerous: where the organization's confidence is running ahead of its evidence.

A hot cell on the coverage matrix can still contain a gap. A product area that has been studied extensively—but two years ago, or studied only at the surface level through comprehension tests that never explored motivation—looks populated but may not support the assumptions the team is currently making. The team thinks it knows something. The evidence supports something narrower. That distance between what the organization believes and what it can actually demonstrate is where the most consequential Frame failures happen. Not in the cold cells, where everyone knows there is nothing, but in the warm cells, where everyone thinks they know but the knowing is softer than it looks.

The assumption audit makes that distance visible. Take a concrete example. The organization believes that small-business owners evaluate pricing by comparing feature tiers. The evidence is a sprint study from eighteen months ago in which twelve participants navigated a pricing page. The study tested whether users understood what each tier included. It did not test how users think about pricing decisions, what criteria they actually use, or whether tier comparison is even how they approach the category. The belief is an inference from a comprehension test. The evidence does not reach the claim. That is a scope gap—one of three forms the gap takes.

The first is a confidence gap. The team believes something strongly. The evidence supports it weakly. A handful of participants in one sprint study two years ago said something directionally consistent with the current belief. That is not the same as strong evidence. The confidence gap asks: if someone challenged this belief in a planning meeting, could you defend it with specific evidence from a well-designed study? If the answer is not clearly yes, the belief is softer than the organization treats it.

The second is a recency gap. The evidence exists and was strong when it was produced, but the product has changed, the market has shifted, or the user population has evolved in ways that may have invalidated the original finding. The recency gap asks: is this evidence still likely to reflect current reality, or are we citing findings from a context that no longer exists?

The third is a scope gap. The evidence is real and recent, but it answers a narrower question than the belief it is being used to support. A comprehension test showing users understood a pricing page is being treated as evidence that users are comfortable with the pricing model. Those are different claims. The scope gap asks: does the evidence actually support the belief, or is the belief an inference the organization has been making from evidence that does not quite reach it?

The assumption audit does not need to cover every belief in the Frame. It is most valuable when run against the assumptions that are about to drive a significant decision. Before a major product bet, before a market expansion, before a strategic pivot, the assumption audit asks: what are we taking as given here, and how well does our evidence actually support it? The cells that light up are the research priorities for the next cycle.

These four representations serve different purposes. The structured document is where the Frame lives. The coverage matrix is how you govern it. The belief map is how you communicate it. The assumption audit is how you stress-test it. A team with one researcher and a focused product area might maintain a lightweight structured document and produce a belief map when needed, with no formal coverage matrix. A large organization with multiple product areas and overlapping user populations needs all four, with the coverage matrix then functioning as the mechanism that makes the federated Frames visible as a system rather than as isolated team artifacts.

The implementation scales to the organization. What does not scale down is the underlying principle: the Frame needs to be explicit, maintained, and owned. Whether it lives in a shared document, a dedicated tool, or a combination of both is a decision each team makes based on its context. The structure described here is a starting point, not a prescription.

Operationalizing the Frame

The definition of the Frame is clear. The harder question is how to make it real inside an actual organization with actual constraints.

Persuasion alone is insufficient.

Research leaders who plan to bring the Frame concept to their leadership with a deck explaining why it matters should lower their expectations for what that will produce. Explaining why the Frame is important generates agreement in the room and inaction in the months that follow. Organizational structures do not change through intellectual persuasion alone. They change when the cost of the current structure becomes visible and attributed.

What actually moves organizations is making the absence of the Frame concrete and expensive at a moment when someone with authority is paying attention. A product decision gets made on two-year-old assumptions about user motivation, and the launch underperforms in a way that current research would have predicted. Three separate sprint studies in a single quarter reconstruct the same foundational context from scratch because nobody knew what the organization already knew, which is a waste that can be calculated in researcher hours. A new feature fails to drive adoption despite strong micro-study results, because the micro studies were testing within a Frame that no longer matched how users actually worked.

When any of these things happen in front of a leader who has budget authority, the conversation shifts from research philosophy to operational risk. That is the moment to have a specific proposal ready: here is what Frame maintenance requires, here is what it costs, here is what the organization is currently paying for the absence of it.

For teams that cannot wait for a visible failure, the tactical approach is to start with the most obviously stale part of the Frame rather than trying to build the whole thing at once. Pick one area where the organizational belief about users is clearly outdated. Make the staleness concrete. Document what decisions are being made against it and what the exposure is. Use that one case to establish the precedent that the Frame is something the organization actively maintains. One concrete demonstration of why Frame maintenance matters is worth considerably more than a comprehensive proposal for why it should exist.

Ownership by whoever cares most is not ownership.

The most common failure mode in Frame maintenance is informal ownership by a highly motivated individual. One researcher—usually a principal or senior lead—takes it upon themselves to maintain the Frame. They track which parts of the organizational understanding are getting stale. They flag when foundational assumptions seem to be drifting from reality. They advocate for deep research work when the Frame needs rebuilding.

This works until it stops. The researcher leaves, gets promoted into a role that consumes all available time, or burns out from maintaining something that has no organizational support and is not reflected in how they are evaluated. Within two quarters, the informal maintenance stops, and the organization is back to archive-plus-implicit-assumptions, with no visible moment of transition.

Real ownership means the Frame is explicitly in someone's job description, with protected time attached—not as part of their broader responsibilities alongside everything else, but as a defined responsibility with protected time that does not get cannibalized when demand for fast work increases. The amount of time required is not large; a principal researcher spending 10 percent of their capacity on Frame maintenance is enough for most organizations in a steady state. What matters is that the 10 percent is real and defended rather than aspirational and sacrificed at the first sign of backlog pressure.

Real ownership also requires the Frame steward to have organizational permission to call for deep research work without going through the standard research request process. Deep research work that is not attached to an immediate product question does not pass standard prioritization criteria: there is no specific decision it is informing, no immediate deadline, no PM who championed it. Getting it approved through normal channels requires a special argument every time, which creates friction that compounds until the Frame steward stops trying. The permission to flag Frame-level risk and initiate deep research as a function of their role—rather than as a negotiation they have to win each time—is what makes ownership operational rather than nominal.

The Frame is a program, not an artifact.

Organizations that attempt to build the Frame by producing a document have already made the most common mistake. Documents go stale by definition. A foundational research synthesis, however thorough, is accurate at the moment it is written and increasingly approximate thereafter. Treating the Frame as something that gets built once and then referenced is how organizations end up with the outdated-but-revered foundational study problem: a document that everyone cites and nobody updates because nobody owns the updating.

A maintained Frame is an ongoing program with four operational components.

The first is regular access to users that is not tied to any product question. Not a large investment, but enough to track how the user population is shifting, what new tensions are emerging, and where the existing model is starting to describe people who no longer quite exist. This requires protected capacity and a standing commitment that it happens on a cadence rather than when someone has bandwidth.

The second is periodic synthesis that asks not "what did we learn this cycle?" but "how has our understanding of the user changed, and what in the Frame needs to be updated?" Standard synthesis documents what was found. Frame synthesis asks what the organization now believes that it did not before, and what it believed before that it now needs to retire. That distinction requires a different analytical posture and produces a different kind of deliverable.

The third is a process for retiring outdated knowledge. Repositories grow in one direction. Nothing is ever removed. The result is an undifferentiated accumulation where a researcher trying to understand what the organization knows must distinguish current beliefs from historical artifacts under time pressure and without guidance. Active retirement of outdated knowledge, with an explicit record of what has been superseded and why, is what separates a maintained Frame from a growing archive that happens to contain some accurate information somewhere.

Project Retros and the Frame

Every study produces two things: the deliverable the stakeholder receives, and the organizational learning that should update what the team believes about its users. Right now, only the first thing happens systematically. The study ends, the readout lands, everyone moves on. The learning stays in the researcher's head or a Confluence page nobody revisits. The Frame does not get updated because there is no moment where anyone asks what changed in what we know.

A project retro creates that moment. Not a long process, just a structured thirty minutes after a study ends and before the next project begins. It comes down to three questions.

What did we find that updates something we previously believed about our users? This is the Frame update question. If the answer is nothing, the study either confirmed what was already known—which is useful to document—or the question was too narrow to surface anything foundational, which is worth noting. If the answer is something, that something goes into the structured document with a note that it supersedes the previous belief and why.

What did we find that we still cannot explain? This is the Frame-level question log question. Participants responded in a way the team did not anticipate. A pattern emerged that falls outside the current Frame. The study was not designed to answer it. These moments are the most valuable outputs of fast research—and the most consistently lost. A retro gives them a home. They go into the question log as candidates for the next deep research cycle.

What would we do differently? This is the methodology question. It does not feed the Frame directly, but it feeds the quality of the work that maintains the Frame over time. Prompt design that did not work; recruitment criteria that produced the wrong participants; a synthesis step that took twice as long as it should have. Small improvements compound across a quarter.

The retro is the fourth operational component of Frame maintenance, alongside regular user access, periodic synthesis, and retiring outdated knowledge. It is the cheapest of the four because the work is already done. You are not running additional studies or scheduling additional sessions. You are adding thirty minutes of structured reflection to work that happened anyway. The return is that organizational learning stops evaporating the moment a readout is delivered and starts accumulating into the Frame, where it can actually be used.

The format does not need to be elaborate. A shared document with the three questions, filled in by the researcher who ran the study and reviewed briefly with the Frame steward, is enough. The discipline is what matters, not the ceremony. A team that does this consistently after every study will find its Frame noticeably more current and more useful within a quarter than a team running the same studies without the retro step.

The Delta

The delta is what a study produces instead of a deck. The readout—the forty-two-slide presentation that gets a meeting and then a Confluence page nobody visits—is the standard output of the event model of research. The delta is the output of the Frame model. Where the readout delivers findings to an audience, the delta updates a state. Where the readout needs to earn attention and persuade stakeholders that the research matters, the delta assumes it matters and gets to the point: here is what the organization believed, here is what the evidence says, here is the gap between those two things, and here is what that means for the decision being made right now.

A delta answers five questions. What does the organization currently believe about this user population? What does the evidence actually say? How large is the distance between those two things? Where is the organization operating on assumption rather than evidence? And how confident are we in the update, given the sample and the method?

The hardest of those five questions is the first one. Articulating what the organization currently believes requires making explicit the assumptions that usually live in a PM's head, in a year-old strategy document, or in something a VP said in planning that everyone absorbed as truth. Making that belief visible and then placing the evidence next to it is where the value lives. The study might confirm the belief; it might contradict it. It might reveal that the belief was never tested and the organization has been operating on a hunch dressed up as institutional knowledge. That distance between belief and evidence is the delta—not the finding, but the gap.

What the Frame Produces

A program still produces things. The Frame is not a document, but it generates documents, and understanding the difference between an artifact produced from the Frame and an artifact that replaces it is what keeps the program from collapsing back into the old model.

Leadership will ask for deliverables. That is not a problem. The problem is when the deliverable becomes the goal rather than a byproduct. A journey map produced from the Frame is useful and shareable. A journey map that gets filed and never updated while the Frame is quietly abandoned is exactly the outdated-but-revered artifact problem described earlier in this chapter.

The practical rule is this: artifacts from the Frame should always trace back to the structured document, not replace it. A belief map produced for a planning cycle is a synthesized view of the Frame at that moment. When the Frame updates, the belief map gets updated or retired. The artifact is a snapshot. The Frame is the source.

Three artifact types work well in practice without threatening the Frame's integrity. The first is the belief map, already described in this chapter, which translates current organizational knowledge into a form leadership can engage with. Produce it at the start of planning cycles; update it when the Frame updates; and label it with a date and a version so everyone knows it reflects the Frame at a specific moment. The second is a gap report: a short document that names the cold and lukewarm cells in the coverage matrix and connects them to upcoming product decisions that carry Frame risk. This gives leadership something concrete and actionable without requiring them to engage with the Frame directly. The third is a Frame update memo: a brief note, typically one to two pages, that is produced after deep research completes and states what the organization believed before, what changed, and why. This is the artifact that retires old beliefs formally and establishes new ones. It makes the program visible without turning it into a static document.

The test for any artifact produced from the Frame is whether it could be mistaken for the Frame itself. If it could, it needs a version date, an owner, and an explicit statement that it is a view of the Frame at a specific point in time—not the Frame itself.

The Frame can become dogma.

If the people who steward the Frame treat ownership as a monopoly on interpretation rather than accountability for accuracy, the Frame stops being a living model and becomes an orthodoxy. The foundational understanding that was supposed to reduce organizational uncertainty starts producing a different kind of uncertainty: nobody can update what the organization believes, even when the evidence has clearly shifted, because the Frame steward defends the existing model rather than maintaining it.

A healthy Frame is stable enough to guide work and contestable enough to revise. The governance structures around it need to include a real process for surfacing disagreements about what is true about users, a defined standard of evidence for changing an organizational belief, and explicit acknowledgment that the Frame steward's job is to keep the model honest—not to defend it.

Building the Frame from Scratch

Most of the guidance in this book assumes the Frame exists in some form, however incomplete or stale. But many teams reading this will not have one. They have research artifacts—studies that were run, decks that were presented, Confluence pages that were filed. What they do not have is a synthesized, actively maintained model of their users that the team actually uses to make decisions. That is a different thing from having done research, and the gap between the two is where most research functions live.

Building the Frame from scratch is deep research work. It is the most important investment a research function can make, and it is almost always the hardest to get funded because it does not produce a two-day readout and is not connected to any single product decision. This section is about how to do it practically, starting from whatever you have.

Start with what already exists.

Before scheduling a single user session, do a research audit. Pull every study, every readout, every synthesis document the team has produced in the past two to three years. Include customer support data, sales call recordings, NPS verbatims—anything that contains direct user signal. The goal is not to synthesize everything into a comprehensive document. The goal is to find out what you already know and, more importantly, where the gaps are.

What you are looking for as you review is not the findings themselves but the pattern across them. What do the studies keep touching without ever fully answering? Where do the findings contradict each other in ways that suggest the team's underlying model of users is wrong or incomplete? What topics have never been studied at all? What is everyone on the team confident about that has never actually been validated?

The audit produces two things: a rough draft of what the Frame currently contains—the user understanding the organization is already operating on, even if implicitly—and a list of the Frame's biggest gaps, the places where the team is building on assumption rather than evidence.

Build the Frame around questions, not findings.

The most common mistake in building a Frame is treating it as a synthesis of existing research. You pull everything you have, organize it by theme, write up summaries, and call it the Frame. This produces a well-organized archive of past studies. It does not produce a Frame.

A Frame is organized around questions about users, not around findings from studies. The questions are the operating categories: how do users in this segment think about the problem this product solves? What is their current workflow, and where does the product fit—or not fit—within it? What do they trust, and what makes them skeptical? What are the stakes for them when something goes wrong? How do they evaluate options in this category?

Each question has an evidence status: answered with strong evidence, answered with weak or dated evidence, partially answered, or unanswered. The Frame is not a document that tells you what you know. It is a map that tells you where your evidence is solid, where it is thin, and where it is missing entirely. That map is what makes the Frame useful for decision-making rather than just interesting to read.

Run the foundational study.

The audit and the question mapping tell you where the Frame's biggest gaps are. The foundational study is how you fill them.

A first Frame-building study is not a micro study and it is not a sprint study; it is deep research. It consists of extended conversations with users—probably ten to fifteen in the first pass—designed to answer the questions the audit identified as most important and most unknown. These are not task-based sessions. They are open, exploratory conversations about how users think about the problem domain, what their actual workflow looks like, what they have tried before, and what they wish existed. The goal is not to validate assumptions. It is to find out what the team does not know it does not know.

The sessions should be researcher-led. Not because AI moderation is inadequate for everything but because the foundational study is exactly the kind of work where the unexpected response is the most valuable data—and AI moderation cannot follow an unexpected thread the way a skilled researcher can. This is the work that justifies the researcher's presence in the organization. It cannot be automated.

Plan for four to six weeks for a first Frame-building study, including synthesis. That timeline will feel slow relative to the micro and sprint work this book describes. It is slow. It is also the investment that makes everything else faster. A team running micro research against a current, accurate Frame is producing signal that connects to reality. A team running micro research against a stale or absent Frame is producing precise answers to questions that do not matter.

Document the Frame as a living artifact, not a final report.

The Frame document is not a research report. It does not have a methods section, an appendix of transcripts, or a limitations paragraph. It is a reference document for the team that builds and ships the product, and it should be written and organized for that audience.

The Frame document should be organized around the questions from the mapping exercise, not around study timelines or research themes. Each section states what the evidence says, how confident the team should be in it, and when it was last validated. It includes the mental models and workflows that emerged from the foundational study, usually as visual artifacts that can be put on a wall or shared in a planning document. It names what is unknown as explicitly as what is known, because the gaps are as important as the content. And it has an owner: a named researcher whose job includes keeping it current.

That last point matters more than it sounds. A Frame document without an owner goes stale immediately. Within six months, it will be out of date. Within a year, it will be actively misleading. The Frame is not a deliverable. It is an ongoing responsibility. The researcher who builds it is not done when the document is written. They are done when the team stops needing it—which is never.

What a minimum viable Frame looks like.

Not every team has four to six weeks for a foundational study. Some teams are under enough pressure that the perfect Frame-building plan becomes the enemy of any Frame at all.

A minimum viable Frame is better than no Frame, and it can be built faster. The minimum version has three things. First, it has a documented answer to the most important foundational question for the product: what is the actual problem users are trying to solve, and how do they currently solve it? This alone, if it is evidence-based rather than assumed, changes the quality of micro and sprint research that follows. Second, it includes a user workflow map for the core use case: what are users doing before they interact with the product, during the interaction, and after? Where does the product fit in their actual workflow, and where does it conflict with it? Third, it includes a gap list: the questions the minimum Frame cannot answer and that the team is currently assuming. This makes the assumptions visible rather than invisible; that visibility is the precondition for eventually testing them.

A minimum viable Frame built in two to three weeks through a focused set of eight to ten exploratory sessions is not as comprehensive as a full foundational study. It is infinitely better than building on nothing, and it gives the team something to pressure-test and improve over time rather than a blank space where the Frame should be.

The Frame does not have to be complete before the engine can run. It has to exist and be treated as something worth maintaining. That is the standard. Start there.

The Frame at Scale

In a large organization where different product areas serve different user populations, a single Frame is the wrong structure. The principle extends to scale, but the operationalization requires adjustment.

What works at scale is a federated structure: distinct Frames for distinct product areas, maintained by the UXR teams embedded in those areas, with shared governance principles that ensure consistency of rigor and a mechanism for identifying when different areas are operating on contradictory beliefs about the same user. Most large organizations have the federated Frames informally. The shared governance layer is what most of them are missing—the structure that makes contradictions between Frames visible and resolvable before they produce conflicting product decisions.

Frame quality at scale is best assessed through the four properties applied at the area level rather than through a single organizational metric. Coverage, freshness, confidence, and ownership are evaluable for each product area independently. The governance question is whether someone has visibility across Frames to identify when they are pulling in different directions.

The Cost of Getting This Wrong

When the Frame is wrong, every answer produced within it is wrong, regardless of how rigorously it was produced. A micro study that carefully tests user comprehension of a pricing page assumes that the pricing model addresses a problem users actually have, that the user segment is correctly identified, and that the context in which users encounter the page matches what the team imagines. Those assumptions are Frame-level claims; if any of them is incorrect, the micro study will produce a technically correct finding that supports a strategically wrong direction.

This is not a hypothetical. It is the most common pattern by which research-informed product development fails—not because the research was executed poorly but because it was executed within a Frame that no longer matched reality. The research was right about what it measured; it just was measuring the wrong things.

Organizations that build and maintain a functioning Frame are not just doing better research. They are building a different kind of research function—one that produces an understanding that compounds over time rather than findings that expire with each readout. The speed and AI tooling described throughout this book enable that function to operate at a pace that was previously impossible. The Frame is what gives that speed direction.

Without it, you are running the engine in neutral. Fast. Loud. Going nowhere in particular.

Where This Goes Next

This chapter defined the Frame: the organization's accumulated, actively maintained model of its users. Not the repository, not the persona set, not the journey map; the living belief system that all three research modes depend on. It introduced the four properties worth assessing—coverage, freshness, confidence, and ownership—and laid out what Frame maintenance actually requires: organizational permission, protected time, a program rather than a document, and governance structures that keep the Frame honest rather than entrenched.

The Frame is the foundation. The next two chapters cover the two fast operating modes that run within it. Chapter 6 covers sprint research—the middle mode for questions that need more than directional signal but less than a foundational investigation. One to two weeks; researcher-led follow-ups on top of AI-assisted data collection. Chapter 7 covers micro research—the fastest mode, for narrow questions with specific decisions attached, running in twenty-four to seventy-two hours when the infrastructure is in place.

Both chapters assume the Frame is functional. When it is not, the routing logic still works, but the outputs become less useful over time in ways that are hard to diagnose from inside the fast work. The Frame is what makes the fast work right, not just fast.

CHAPTER 6

Sprint Research

Sprint research is the middle mode. It is for questions that are too broad or too ambiguous for a twenty-four- to seventy-two-hour micro study but too focused and time-sensitive for a multi-month foundational investigation. The domain is generally understood. The question is real but not fully shaped. The stakes are meaningful enough that directional signal alone is not sufficient. You need something with more depth, more evidence, and more interpretive weight—but you need it in one to two weeks, not one to two months.

This is also the mode that is hardest to describe cleanly, which is part of why it deserves its own chapter. Micro research has a crisp definition: one question, one decision, fast turnaround. Deep research has a familiar shape: open-ended, extended, exploratory. Sprint research sits between them, and that middle position makes it feel less defined. Some teams call it rapid research. Some call it focused research. Some do not name it at all but recognize the pattern when you describe it: a study that moves fast because AI tools compress the data collection, but goes deeper than micro because the researcher stays involved throughout, conducts follow-up sessions, and synthesizes across a broader evidence base.

This chapter defines the sprint research operating mode: when to use it, how to scope it, how to structure AI-first data collection with researcher-led follow-ups, how to analyze with AI support while maintaining evidence integrity, and how to deliver findings in a format that matches the timeline.

When Sprint Research Is the Right Mode

If you have internalized the routing logic from Chapter 4, you already know the general rule: sprint research fits when the routing dimensions point to the middle of the spectrum, where there is moderate risk (the decision matters enough that a quick directional answer is not sufficient, but it is not high-stakes enough to justify a six-week foundational study); moderate ambiguity (the question is understood well enough to design a study, but the answer is not obvious and may require follow-up probing that AI moderation cannot handle alone); and medium expiry (the findings will be relevant for weeks or months, not days, which justifies the additional investment of researcher time).

Concretely, sprint research fits questions like these: How do users in a specific segment experience a multistep workflow, and where does it break down? What are the main objections to a new-feature concept, and how do they vary across user types? How do users understand and navigate a pricing structure that has changed? What mental models do users bring to a product category that the team is entering for the first time? These are not questions that can be answered by showing someone a screen and asking if they understand it. They require conversation, follow-up, and the kind of interpretive judgment that comes from a researcher hearing something unexpected and knowing to dig deeper.

A useful heuristic: if the question could be answered by a well-designed AI-moderated study alone, it is probably micro research. If answering the question well requires a researcher to hear something, react to it, and follow up in real time, it is sprint research. The need for researcher-led follow-up is what distinguishes the two modes at a practical level. Micro research is AI-moderated with researcher oversight. Sprint research is AI-assisted with researcher involvement. That distinction sounds subtle. It is not.

The Shape of a Sprint Study

Sprint research has a recognizable structure, though the specifics vary by question. The general pattern is AI-first data collection at the front, researcher-led follow-ups in the middle, and AI-assisted analysis at the back—with the researcher controlling interpretation throughout.

The front end is AI-moderated sessions with real participants, and this is where the tooling dependency is most visible. Using tools such as Outset, the researcher designs the study, writes the prompts and probes, sets the guardrails, and launches. Participants respond asynchronously via text, audio, or video. The AI follows the script, asks follow-up probes based on responses, and captures everything with full transcripts, timestamps, and linked quotes. This phase can run in parallel with ten, fifteen, or twenty participants, compressing what would take a week or more with researcher-moderated scheduling into one to three days. That compression is not a convenience. It is what makes the two-layer structure of sprint research structurally possible. In a traditional model, collecting data from fifteen to twenty participants would consume the entire one- to two-week timeline, leaving no room for the follow-ups that give sprint research its depth.

The volume matters. In a traditional sprint-length study, a researcher might conduct eight to twelve sessions over a week, limited by scheduling, energy, and the cognitive load of running live interviews back to back. With AI-moderated collection, you can have responses from fifteen to twenty participants within days. That broader base means you enter the follow-up phase with a richer initial dataset, more variation in perspectives, and a better sense of which threads are worth pulling.

The middle is where the researcher earns their keep. After reviewing the AI-moderated responses, the researcher identifies participants whose answers were surprising, ambiguous, or particularly rich, then selects four to six of these participants for follow-up sessions that the researcher conducts directly. These are not full repeat interviews. They are targeted follow-ups—ten to twenty minutes, often less—focused on specific themes or moments that emerged from the initial data and that require human probing to understand.

This is the step that separates sprint research from micro research. In micro research, the AI-moderated responses are the dataset. The researcher reviews, audits, and interprets, but does not go back to participants. In sprint research, the AI-moderated responses are the first layer. The researcher-led follow-ups are the second layer. The combination produces something that neither layer could produce alone: breadth from the AI-moderated sessions and depth from the researcher-led ones.

The back end is synthesis and delivery, which I will cover in detail later in this chapter.

Scoping Sprint Research

The scoping discipline for sprint research is different from micro research. Micro research scope is ruthlessly narrow: one question, one decision. Sprint research scope is broader but still bounded. You need to define the territory the study will cover without letting it expand into open-ended deep research.

A useful scoping approach is to Frame the sprint study around a central question with two to four sub-questions that represent the specific angles the research will investigate. The central question provides direction. The sub-questions provide structure. Together, they give the study enough room to surface unexpected findings without drifting into formless exploration.

For example, a sprint study might have this as a central question: how do small-business owners evaluate and choose between pricing tiers for our product? The sub-questions might be: What information do they need to make the decision? What is confusing or concerning about the current pricing page? How does this decision fit into their broader purchasing workflow? What would cause them to choose a lower tier than expected or abandon the process entirely?

Each sub-question can be addressed through the AI-moderated sessions. The follow-up sessions then dig into the themes that emerge across sub-questions, the contradictions between participants, and the moments where the AI-moderated data raised more questions than it answered.

The scoping trap for sprint research is letting it become deep research by another name. If the central question is genuinely open-ended, the study will drift because there is no anchor. "What do small-business owners need from our product?" is not a sprint research question. It is a deep research question wearing a sprint timeline. Sprint research works when the territory is defined and the study explores within it. It does not work when the territory itself is unknown.

Another scoping discipline worth building into your practice is defining what the study will not cover. This sounds obvious, but it is frequently skipped. A sprint study on pricing evaluation should explicitly state that it will not assess overall product-market fit, that it will not compare the product to competitors, and that it will not investigate long-term retention behavior. These boundaries prevent stakeholders from loading additional questions onto the study mid-flight, which is how sprint studies balloon into unmanageable projects. If you read Chapter 4 and thought the decision contract was overkill for fast work, sprint research is exactly where it pays for itself. The boundaries section of the contract is what keeps a two-week study from becoming a six-week study.

In practice, sprint research requests rarely arrive scoped. A PM says, "We need to understand how users feel about the new pricing."" That is a topic, not a study. The scoping work is turning that topic into a central question and sub-questions that a two-week study can actually answer.

Start by asking what decision the research is informing. The PM says the team is deciding whether to ship the new pricing page as-is or revise it before launch. Now you have a decision. Next, ask what specifically they are uncertain about. The PM says they are not sure if users understand the tier differences, whether the comparison layout works, and whether the annual discount is compelling enough to shift behavior. Those are three sub-questions, and each is testable.

The central question becomes: how do target users evaluate and make decisions on the new pricing page? The sub-questions become: Do users understand what differentiates the tiers? Does the comparison layout help or hinder the evaluation process? And how do users respond to the annual discount framing? Each sub-question can be addressed in the AI-moderated phase. The follow-ups dig into the reasoning behind the patterns.

Notice what got excluded. The PM's original framing—"how users feel about the new pricing"—could have spiraled into competitive pricing comparisons, willingness-to-pay analysis, or general sentiment about the product's value. None of that fits a two-week sprint. The scoping conversation drew a boundary around what the study will cover and, just as importantly, what it will not. That boundary is what makes the timeline achievable.

Designing the AI-Moderated Phase

The AI-moderated phase of a sprint study uses the same tools and many of the same principles as micro research, but the prompt design reflects the broader scope. Where a micro study might have five to eight tightly focused prompts, a sprint study might have twelve to eighteen prompts organized around the sub-questions.

The prompts should still be specific and concrete. The difference is that sprint research prompts can include more open-ended follow-up probes, because the AI-moderated data is not the final dataset. It is the first layer. If an open-ended probe produces thin or ambiguous responses, you can address that directly in the follow-up sessions. This gives the prompt design slightly more room to breathe than in micro research, where the AI-moderated data is all you have.

Stimulus design matters just as much in sprint research as in micro. Wherever possible, prompts should be anchored in concrete artifacts—screens, flows, prototypes, copy, pricing tables, onboarding sequences. Asking participants to react to something real produces better data than asking them to speculate about hypotheticals. This is true for both the AI-moderated phase and the follow-up sessions.

A design pattern I have found effective is to use the AI-moderated phase for tasks and reactions, and reserve the follow-up sessions for sense-making conversations. The AI handles "what do you think this means" and "where do you get stuck" efficiently. The researcher handles "why did that confuse you" and "tell me more about how you think about this" better. Dividing the labor this way plays to the strengths of each modality. The AI is good at consistent, parallel execution of structured tasks. The researcher is good at listening, reacting, and following threads that the script did not anticipate.

One more design consideration: the AI-moderated phase should be structured so that reviewing responses afterward makes participant selection for follow-ups straightforward. This means the prompts need to be designed not just for the participant's benefit but also for the researcher's review. If the prompts are too open-ended, the responses will be sprawling and hard to compare across participants. If they are too closed, the responses will be thin, and there will not be enough signal to identify which participants to follow up with. The sweet spot is prompts that are specific enough to produce comparable responses but open enough that participants with interesting perspectives reveal themselves.

Before moving to follow-ups, there is a decision point that deserves its own moment: the go/no-go assessment after the AI-moderated data comes back. This is the sprint research equivalent of the first-three-participant audit in micro research, but it covers the full AI-moderated dataset.

Review the transcripts, not just the summaries. Ask three questions. First, did the prompts work? In other words, are participants responding to the actual questions, or are they misunderstanding, going off-topic, or giving thin responses that suggest a problem in the study design? If the prompts failed, the follow-up sessions cannot fix that. You need to decide whether to revise and rerun the AI phase or proceed with what you have.

Second, is the question actually sprint-level? Sometimes the AI-moderated data reveals that the question is simpler than expected. Fifteen participants all understood the pricing page, the confusion is isolated to one element, and a micro-level readout would answer the question. Running follow-up sessions on a question that is already answered wastes time and participant goodwill. If the AI-moderated data answered the question, write the readout and move on. You scoped a sprint and got a micro result. That is not a failure. That is efficiency.

Third, is the question bigger than expected? If the AI-moderated data surfaces fundamental confusion, conflicting mental models, or problems that extend well beyond the study's sub-questions, you may be looking at a deep-research-level question that got misrouted. The decision contract helps here. If the emerging findings fall outside the boundaries you defined, flag it. Deliver what the sprint study can support, and recommend a deep research project for the rest. Do not try to stretch a two-week sprint to answer a question that needs two months. That is how sprint studies produce findings that are technically complete and practically misleading.

Selecting and Conducting Follow-Up Sessions

The follow-up sessions are the highest-value step in sprint research and the one that requires the most researcher judgment. The goal is not to reinterview every participant. It is to select the participants whose responses in the AI-moderated phase were most informative, most surprising, or most ambiguous—and dig deeper.

Selection criteria matter. There are several useful lenses:

Divergence: Participants whose responses differed significantly from the majority, because understanding outliers often reveals the most about the problem space

Richness: Participants who gave detailed, nuanced responses that suggested they have more to share than the AI-moderated format can capture

Ambiguity: Participants whose responses were unclear or contradictory, because resolving that ambiguity through conversation often produces the most useful findings

You will not use all three criteria for every study. The right lens depends on what the AI-moderated data showed you.

The follow-up sessions themselves should be focused, not free-form. A ten- to twenty-minute session is typically sufficient. You come in with specific moments from the participant's AI-moderated responses to probe: "You mentioned that the pricing page was confusing. Can you walk me through what you were looking at and what was unclear?" Or "You chose option B but said you were not confident. What would have made you more confident?" Or "You said you would need to check with someone else before deciding. Who would that be, and what would they need to know?"

These are not cold interviews. You have already seen the participant's responses; the participant knows you have seen them. This creates a different dynamic than a traditional first-contact interview. The conversation can start from a shared baseline and go deeper immediately. I have found that follow-up sessions are consistently more efficient and more productive than first-contact sessions because both parties are already oriented. You are not spending the first ten minutes on context-setting. You are spending them on the thing that actually matters.

A practical note on logistics, because this is where sprint research breaks down if you do not plan ahead. The follow-up sessions require getting participants back on a live call two to three days after their AI-moderated session. That does not happen by accident. You need to build the expectation into recruitment. When recruiting, tell participants that the study has two parts: an asynchronous session they complete on their own time and a possible short, live follow-up a few days later. Not every participant will be called back, but all should be available. Screen for participants who can commit to both. Offer appropriate incentives for both parts. If you recruit without setting this expectation, you will finish the AI-moderated phase, identify the six participants you want to follow up with, and discover that three of them are unavailable. You have lost your best follow-up candidates, and the sprint timeline does not allow you to recruit replacements.

For this reason, I suggest overrecruiting slightly. If you need four to six follow-ups, recruit eighteen to twenty for the AI-moderated phase, knowing that not all will be available for the second part. The surplus also gives you a richer initial dataset, which is a side benefit.

The number of follow-ups depends on the study scope and the patterns in the AI-moderated data. Four to six is typical. Fewer than three risks missing important nuance. More than eight starts to turn the sprint into a traditional study, which defeats the purpose. The follow-ups are strategic samples from a larger dataset, not an attempt to talk to everyone.

Analysis with AI Support

Sprint research produces a larger dataset than micro research: fifteen to twenty AI-moderated transcripts plus four to six follow-up session recordings. Analyzing this volume in a one- to two-week timeline requires AI support, but the same cautions from Chapter 3 apply. AI-generated summaries are first drafts. They are not findings.

The analysis workflow I recommend is a three-step process. First, use AI tools to generate initial summaries and identify patterns across the AI-moderated transcripts. These summaries highlight where participants converged, where they diverged, and which themes appeared most frequently. Tools such as Outset link summary claims back to specific quotes with timestamps, which makes this first pass faster and more traceable than working with raw transcripts alone. Second, review the summaries against the raw transcripts, checking for the failure modes described in Chapter 3: smoothing away contradiction, inflating agreement, missing quiet but important signals. Third, integrate the follow-up session data, which you analyze more carefully because the depth and nuance of researcher-led conversations do not compress well into automated summaries.

The integration step is where the real synthesis happens. The AI-moderated data provides the breadth—patterns across fifteen to twenty participants. The follow-up data provides the depth—the why behind the what, the context behind the behavior, the explanation behind the confusion. A sprint research finding typically rests on both layers: a pattern observed across the broad dataset, enriched and explained by specific moments from the follow-up conversations.

Evidence chains are essential. Every claim in the sprint research deliverable should be traceable to specific participant moments, with timestamps and quotes. This is not optional rigor for the sake of academic credibility. It is practical protection. When a stakeholder questions a finding, you need to pull up the supporting evidence in under a minute. When a finding gets cited in a road map discussion three months later, someone needs to be able to verify what it was actually based on. The evidence chain is what makes that possible.

One analytical discipline that I think is underappreciated in fast research is the disconfirming evidence standard. In micro research, the dataset is small enough that contradictions are easy to spot. In sprint research, with a larger dataset, it becomes tempting to let the majority pattern dominate and treat contradictions as noise. That temptation should be resisted. Two participants out of eighteen who describe a fundamentally different experience might represent a segment the team has not considered—or they might be outliers. You do not know, and the sprint study probably cannot tell you which. But what you can do is flag them. Present the majority pattern and then present the contradiction. Let the decision maker weigh it. Do not bury it because it complicates the narrative.

Delivering Sprint Research Findings

Sprint research deliverables need to match the timeline and the audience. A two-week study does not justify a forty-slide deck. It also should not be a Slack message. The format depends on the organizational culture, but the principles are consistent.

The deliverable should lead with the decision implications, not the method. It should state findings in plain language before providing supporting evidence. It should include explicit boundaries on what the findings can and cannot support. And it should be short enough that the decision maker will actually read it.

Two formats work well for sprint research. The first is a short written report, typically three to five pages, structured around the sub-questions with findings, supporting evidence, and limitations for each. This format works well for teams that need a reference document and for decisions that will be revisited over the following weeks. The second is a focused deck, typically eight to twelve slides, structured around findings with embedded participant quotes and clips. This format works well for teams that make decisions in meetings and need something presentable.

Regardless of format, sprint research deliverables should include several elements: a clear statement of the central question and sub-questions; the key findings, stated as observations rather than recommendations unless you have been explicitly asked for them; supporting evidence linked to specific participant moments; disconfirming evidence or notable contradictions; explicit boundaries on generalizability and confidence; and an expiry date or condition. These findings are valid until the feature changes, the market shifts, or a specified time passes.

A deliverable pattern I have found effective is a two-layer structure. The top layer is a one-page executive summary for the decision maker: findings, implications, limitations—done. The bottom layer is a full evidence appendix that can be audited if anyone questions a finding. The executive reads one page. If they push back on something, you pull up the evidence in thirty seconds. Fast to consume, fast to defend. That is the system.

The expiry condition deserves specific attention because sprint research findings are tempting to reuse. A sprint study that took two weeks of effort and produced substantive findings feels like it should have lasting value. And it does—within its boundaries. But the boundaries matter. A sprint study on how users evaluate a pricing page is valid until the pricing changes. A sprint study on onboarding friction is valid until the onboarding flow is redesigned. Stating the expiry condition explicitly in the deliverable prevents findings from being cited in road map discussions six months later as if they reflect current reality. They do not. They reflect the reality at the time of the study, and that reality has a shelf life.

Common Pitfalls in Sprint Research

Sprint research has its own failure modes, distinct from micro research and deep research. Knowing them in advance does not prevent them entirely, but it helps you recognize them more quickly when they start to happen.

The most common pitfall is scope inflation. A sprint study starts with a clear central question and three sub-questions. During the AI-moderated phase, an interesting tangent emerges; you add it to the follow-up protocol. Another tangent emerges; that gets added too. By the time the follow-up sessions happen, you are trying to answer seven questions instead of three, the sessions run long, and the synthesis becomes unwieldy. The one- to two-week timeline slips. The deliverable arrives late and unfocused.

Scope inflation is the natural consequence of curiosity meeting a richer-than-expected dataset, and the discipline to resist it is one of the hardest parts of sprint research. The mitigation is simple to describe and hard to execute: when a tangent emerges, note it, do not chase it. If it is genuinely important, it becomes the input for a future study. If it is interesting but not relevant to the decision the sprint is informing, it goes into a parking lot. The boundaries you defined during scoping are the anchor. If a tangent falls outside the boundaries, it is out of scope regardless of how interesting it is.

A second pitfall is overreliance on the AI-moderated data and underinvestment in follow-ups. The AI-moderated phase produces a lot of data quickly, and the volume creates a sense of completeness. It is tempting to conclude that eighteen AI-moderated transcripts are sufficient and skip or shorten the follow-up sessions. This produces a study that has the breadth of sprint research but the depth of micro research. You have paid the cost of a sprint study and gotten micro research results. The follow-ups are where sprint research earns its additional investment. Cutting them turns a sprint study into an expensive micro study.

A third pitfall is treating the AI-moderated phase as a screening mechanism rather than a data collection mechanism. If you use the AI-moderated sessions only to identify interesting participants for follow-ups, without analyzing the AI-moderated data as evidence in its own right, you are wasting the breadth advantage. The AI-moderated data should be analyzed, synthesized, and included in the findings. The follow-ups deepen it. They do not replace it.

A fourth pitfall—and this one is more organizational than methodological—is letting sprint research become the default mode for everything. Sprint research takes more time and more researcher involvement than micro research. If every question gets a one- to two-week sprint study, you are back to the capacity constraints of the old model, just with better tools. The taxonomy from Chapter 4 exists precisely to prevent this. Questions that fit micro research should get micro research. Sprint research is for questions that genuinely need the additional depth and researcher-led follow-ups. Using sprint as the default is as much a misclassification error as cramming deep research into micro research.

Where This Goes Next

Sprint research is the workhorse mode for questions that need more than a quick directional answer but less than a foundational investigation. It combines the speed advantage of AI-assisted data collection with the interpretive advantage of researcher-led follow-ups. It produces findings that are broader and deeper than micro research, delivered in one to two weeks instead of one to two months.

But most product decisions that benefit from user evidence are not sprint-level questions. They are narrow, specific, and time-sensitive. They need the twenty-four- to seventy-two-hour operating mode. That is micro research, and it is what the next chapter covers: prompt design, modality selection, quality gates, and the directional readout format that makes fast findings usable without being misused.

CHAPTER 7

Micro Research

Micro research is the fastest operating mode in the taxonomy. It is for narrow, specific, near-term product questions where the decision is defined, the domain is understood, the ambiguity is low, and the question can be answered with real participants in twenty-four to seventy-two hours. The output is a directional conclusion, with explicit boundaries on what it can and cannot support.

If sprint research is the workhorse for medium-complexity questions, micro research is the everyday tool for the dozens of small decisions that would otherwise go unanswered or get decided by whoever talks loudest in the meeting. Does this screen make sense? Is this flow confusing? Which of these two options do users prefer, and why? What is the first objection someone raises when they see this? These are questions that matter, that have real answers, and that can be resolved quickly if you scope them correctly.

Micro research exists because the cost of waiting is often higher than the cost of being directionally right. Not perfectly right—directionally right. There is a difference, and the entire operating discipline of micro research is built around that difference. You get useful signal fast. You state explicitly what that signal can and cannot support. You let the decision maker act on it or not. And you move on.

This chapter covers the full micro research workflow, and it mirrors the sprint research chapter deliberately. The modes are different in scope, depth, and researcher involvement, but the operational logic is the same: define what you are doing, design the study, run it, analyze it, deliver it, and know where the pitfalls are. Chapter 4 defined when to use micro research. This chapter covers how.

When Micro Research Is the Right Mode

The routing logic from Chapter 4 gives you the abstract answer: low risk, low ambiguity, short expiry. The question is clear, the stakes are modest, and the answer is needed fast. But the abstract answer is less useful than a practical heuristic, so here is one: if the question could be answered by a well-designed AI-moderated study alone, without researcher-led follow-ups, it is micro research.

That is the key distinction from sprint research. In sprint research, the researcher needs to hear something, react to it, and follow up in real time. In micro research, the AI-moderated data is the dataset. The researcher designs the study, audits the output, and controls the interpretation, but they do not go back to participants. If the question requires that second layer of human probing, it is not micro. Route it to sprint.

The kinds of questions that fit micro research have a recognizable shape:

Comprehension: What do users think this screen means?

Friction: Where do they get stuck in a short flow?

Clarity: What is confusing about this offer or pricing explanation?

Preference: Given two concrete options with clear trade-offs, which do users choose, and why?

Objection mining: What is the first reason they would not do this?

Terminology: What words land, and what words backfire?

All of these are specific, concrete, and testable against real artifacts. If you can show someone a thing and get a useful answer in five minutes, it probably fits.

Notice the pattern. The constraint is not the number of questions. You can have ten or fifteen questions in a micro research study. That is fine. The constraint is that every question needs to be specific and clear, operating in a well-defined space. Vague questions do not become useful just because you asked a bunch of them quickly. If someone asks you to run a quick study on "what do users want from our product," you do not have a micro research opportunity. You have a scope problem.

The Shape of a Micro Study

Micro research has a simpler structure than sprint research because it has fewer phases. The general pattern is question hygiene at the front, AI-moderated data collection in the middle, and researcher-controlled synthesis at the back. There are no follow-up sessions. The AI-moderated responses are the dataset. The researcher's involvement is concentrated at the beginning and the end—designing the study and interpreting the results.

The front end is question hygiene, which I will cover in detail in the next section. This is where the study is defined, scoped, and connected to a decision. It is the most important step, and in micro research it is even more important than in sprint research because there is no second pass. In a sprint study, if the scoping is slightly off, the researcher can correct course during follow-up sessions. In a micro study, what you designed is what you get. The prompts, the stimuli, the probes. If they are wrong, the data is wrong, and you will not discover that until synthesis, when it is too late to fix.

The middle is AI-moderated sessions with real participants. Using tools such as Outset, you launch the study, and participants respond asynchronously. The AI follows the script, asks follow-up probes based on responses, and captures everything with transcripts, timestamps, and linked quotes. This phase runs fast, often completing within a day once participants are recruited. The volume is smaller than sprint research: eight to fifteen participants is typical—enough to see patterns and enough diversity to know whether the patterns hold across different user types.

One design decision that matters more than most practitioners expect is modality: whether participants respond via text, audio, or video. The tools support all three, and the choice is not neutral.

Text is the fastest to collect and the easiest to synthesize. Participants can respond on their own time with minimal friction. But text compresses signal. You lose tone, hesitation, facial reactions, and the difference between a confident answer and a reluctant one. For straightforward comprehension or terminology questions where the answer is essentially factual, text works fine.

Video captures the most signal. You can see when a participant squints at a screen, hesitates before answering, or says "yeah, that makes sense" while looking confused. For questions where emotional reaction, confusion, or confidence level matters, video is worth the additional friction. It also makes the evidence chain more compelling when presenting findings to stakeholders. A transcript that says "participant was confused" is less persuasive than a fifteen-second clip showing the confusion.

Audio falls between the two. You get tone and pacing without the friction of video. Audio works well when you need more signal than text provides but video feels like too much to ask for a short study.

The general principle is simple: match the modality to what you need to observe. If the question is "do they understand what this means," text is usually sufficient. If the question is "how do they react when they see the price," video captures something text cannot. Default to text for speed. Upgrade to video when the question depends on reaction, not just response.

The speed of micro research is not the result of cutting corners or reducing rigor. It is the result of AI-moderated sessions running asynchronously with multiple participants simultaneously. A researcher moderating sessions manually can run three to four per day at most. An AI-moderated study can collect responses from ten to fifteen participants in a single day because each session runs independently, on the participant's schedule, with the AI handling the moderation. That parallelism is what compresses the timeline from weeks to hours. Remove the tool and the timeline reverts to a traditional cadence, regardless of how tightly the question is scoped.

The back end is synthesis. The researcher reviews the AI-generated summaries, audits them against raw transcripts, applies quality gates, and produces the directional readout. This phase should also be fast. If synthesis takes more than a few hours for a micro study, either the study was not scoped tightly enough or the researcher is overproducing the deliverable.

From start to finish, the entire cycle should take twenty-four to seventy-two hours: question hygiene and prompt design on day one; data collection on day one or two; synthesis and readout by day two or three. If the timeline stretches beyond that, one of two things has happened: either the question was not micro and should have been routed to sprint, or the execution pipeline has friction that needs to be addressed. Micro research earns its value by being fast. If it is not fast, it is just small.

Question Hygiene

The single most important step in micro research happens before any participant sees anything. It is question hygiene: the discipline of making sure the question is actually answerable, actually connected to a decision, and actually scoped to what micro research can deliver.

One question, one decision, one owner. If you cannot name the person who will act on this and what action they will take, stop. You are not ready. This is not a bureaucratic requirement. It is a quality gate. A study with no decision owner produces findings that float. Nobody acts on them because nobody was supposed to act on them. The study existed because someone thought it would be nice to know. "Nice to know" is not a research question. It is a wish.

In practice, no request arrives clean. A product manager says, "Can we test the new checkout flow?" That is not a question. It is a topic. The hygiene process is turning that topic into something a study can answer. What about the checkout flow? Are the team members worried that users will not understand the new step they added? Are they unsure whether the progress indicator is clear? Do they want to know if the shipping options are confusing? Each of those is a different study with a different design. "Test the checkout flow" is none of them.

The move is to ask three questions back: 1) What decision are you making? 2) What are you unsure about? And 3) What would you do differently if the answer surprised you? Most of the time, question 3 does the real work. If the PM says, "Honestly, we are shipping it either way; I just want to know if it makes sense," you now know the actual question: do users comprehend the new flow? That is testable. That is scoped. That is micro research. But you only got there because you refused to accept the first version of the request.

Sometimes the hygiene process reveals that the request is not one question but three. "Test the checkout flow" might mean the team is uncertain about the new step, the shipping options, and the confirmation screen. That is fine. Three tight questions in one micro study is workable. Three vague questions are not. The hygiene step is not about reducing the number of questions. It is about making each one specific enough that you know what a useful answer looks like before a single participant sees anything.

And sometimes the hygiene process reveals that the question is not micro at all. If the PM says, "We want to know if this new flow actually works better than the old one," that is a comparative question with behavioral implications that a five-minute AI-moderated session cannot answer. That is sprint research—possibly with a different methodology entirely. The hygiene step caught it. Better to reroute now than to produce a readout that cannot support the conclusion the team actually needs.

Once you have the question, the decision, and the owner, define three things. First, what will change based on possible outcomes?

If the answer is "we will ship option A" or "we will revise the copy," you have a real decision. If the answer is "we will present the findings," you probably do not. Second, what would change your mind? If no possible finding would alter the team's direction, the study is validation theater and should not run. Third, what will you not conclude from this study? Scope out the conclusions you are not licensed to draw. A comprehension test on a pricing page does not tell you whether users will convert. A preference test between two layouts does not tell you whether the feature solves a real problem. Stating these boundaries before the study starts prevents the findings from being stretched beyond what they can support.

Question hygiene is the micro research equivalent of the decision contract from Chapter 4. For sprint research and above, I recommend writing a formal contract. For micro research, the hygiene check can be faster and less formal, but the same elements need to be addressed: the question, the decision, the owner, the outcomes, and the boundaries. Five things. Five minutes. Skip any of them, and the study is at risk of producing findings that are technically correct and practically useless.

Prompt Design

The execution bottleneck in micro research has moved. It used to be scheduling, moderating, and transcribing. With AI-moderated tools, those steps compress dramatically. The bottleneck is now prompt design. The quality of the questions you put in front of participants determines the quality of everything that comes out.

Keep prompts concrete. Avoid hypothetical future scenarios when you can test comprehension on real artifacts. "Would you use this feature?" is a bad question. It asks participants to predict their own future behavior, which they are reliably terrible at. "What does this button do?" is a better question. It tests comprehension against something specific. "Walk me through what you would do next on this screen" is better still. It combines comprehension with behavioral intention in a way that is grounded in the actual artifact.

Use a small number of tasks or stimuli. A micro study does not need twenty screens. It needs the three or four moments that the team is actually uncertain about. If the team is not uncertain about a screen, do not test it. Testing things you already know the answer to is not thoroughness. It is a waste of participant time and your own.

Use structured probes that reduce moderator drift. In AI-moderated studies, this means writing probes that are specific enough that the AI asks consistent follow-ups across participants. You want comparable data, not a different conversation every time. If participant 3 gets asked about their emotional reaction to the pricing page and participant 7 is asked about prior experience with similar products, you cannot synthesize across them. The probes need to produce data that you can compare.

The prompt design discipline in micro research is tighter than in sprint research because there is no second pass. In a sprint study, if an open-ended probe produces thin responses, you address it in the follow-up sessions. In a micro study, thin responses stay thin. There is no researcher-led recovery. This means the prompts must be specific enough to produce useful data on the first attempt. The margin for error is smaller, which is why the margin for vagueness should be zero.

One prompt design principle is worth repeating: test the thing, not the idea of the thing. Show participants the actual screen, the actual copy, the actual flow. If the artifact is not ready, show the closest approximation you have. Descriptions of features produce opinions about descriptions. Artifacts produce reactions to artifacts. Those are different kinds of data, and only the second kind is useful for the decisions micro research typically informs.

A few concrete examples make the pattern clear.

For comprehension: "Do you understand this pricing page?" is a bad prompt. Participants will say yes because saying no feels like admitting failure. "Look at this screen for thirty seconds. Without scrolling back up, tell me what the three pricing tiers include and how they differ." That tests actual comprehension against the artifact. The participant either retained the information or did not. No self-assessment required.

For preference: "Which of these two designs do you prefer?" is thin. You will get an answer, but not a usable one. "Look at both options. Which one would you use to complete [specific task]? Walk me through your reasons for choosing that option." Anchoring the preference in a task produces reasoning you can analyze. Unanchored preference produces opinions with no explanatory power.

For objection mining: "Would you sign up for this service?" invites speculation. "You have just seen what this service offers and what it costs. What is the first reason you might decide not to sign up?" The framing assumes a reason exists, which gives participants permission to be critical. Prompts that ask "would you" get polite answers. Prompts that ask "why would you not" get honest ones.

For terminology: "Is this label clear?" will almost always get a yes. "Read this label. What do you think happens when you tap it?" tests whether the label communicates what the team intends. If the participant describes something different from what the button actually does, the label failed. You did not need them to tell you it was unclear. You saw it in their answer.

The pattern across all of these is consistent: do not ask participants to evaluate. Ask them to do something, and let the evaluation emerge from what they do. Self-reported clarity, preference, and intent are unreliable. Demonstrated comprehension, task-anchored reasoning, and behavioral responses are usable. Design every prompt around the second category.

Running the Study

In sprint research, the middle phase involves the researcher conducting follow-up sessions. In micro research, there are no follow-ups. The AI-moderated sessions are the data collection, start to finish. But that does not mean the researcher disappears during execution. The researcher's job during the study is monitoring, not moderating.

The most important monitoring step is the first-three-participant audit. After the first three responses come in, stop and review them—not the AI-generated summaries, but the actual responses. Are participants understanding the prompts? Are the probes producing the kind of data you need? Are participants engaging with the stimuli or skipping past them? Are the responses specific enough to support analysis, or are they generic and thin?

This audit is your last chance to catch design problems before the rest of the data comes in. If participants are misunderstanding a prompt, you can revise it before the next ten participants see it.

If a probe is producing useless data, you can replace it. If participants are not engaging with a stimulus, you can adjust how it is presented. Making these corrections after three participants is a minor adjustment. Discovering the same problems after all fifteen participants have responded is a wasted study.

Beyond the first-three audit, monitor for participant quality. In micro research—especially when using panel recruitment rather than your own users—participant quality is a real risk. Are participants giving thoughtful responses or rushing through? Are their responses consistent with the screener criteria? Are there signs of fraud, such as responses that are clearly AI-generated or that contradict basic screening information? The tools are building detection mechanisms for these issues, but the researcher still owns the quality gate. If a participant's data is suspect, exclude it before synthesis rather than letting it contaminate the findings.

Recruitment source matters. Using your own users, recruited from your product, produces higher-quality data than sourcing from general research panels. Your users have real context, real experience, and real stakes. Panel participants have none of those things. Panel recruitment is sometimes necessary for speed or for reaching audiences you do not have access to, but the quality trade-off is real and should be stated in the readout.

Synthesis and Quality Gates

Micro research synthesis is fast by design, but fast does not mean unchecked. The AI-generated summaries are a draft. They are not findings. The researcher's job in synthesis is to audit the draft, apply quality gates, and produce a readout that is evidence-based rather than summary-based.

The quality gates I use are straightforward. Never ship conclusions without checking raw transcripts or clips. The summary says "most participants understood the pricing page." Did they? Pull up the transcripts. Read what they actually said. If the summary is accurate, good. If it smoothed away a participant who was confused but whose confusion was not the majority pattern, you need to include it.

Require at least one disconfirming example per theme. If your finding is "users understood the pricing page," find the participant who did not. If your finding is "users preferred option A," find the participant who preferred option B and understand why. If there is no disconfirming evidence, either you got lucky with a genuinely unanimous finding or you are not looking hard enough. The disconfirming example is not there to undermine your finding. It is there to bound it. "Most participants understood the pricing page. One did not, and the confusion was related to the annual versus monthly toggle" is a better finding than "users understood the pricing page" because it gives the team something specific to act on.

Separate frequency from importance. Five participants mentioning the same minor annoyance does not automatically make it the most important finding. One participant describing a workflow-breaking problem that nobody else encountered might matter more for the decision at hand. Frequency tells you how common something is; importance tells you how much it matters. They are different dimensions, and micro research synthesis needs to track both.

Keep a clear chain from evidence to conclusion. Every claim in the readout should be traceable to specific participant moments. The tools help here; Outset links summary claims back to specific quotes with timestamps, which makes building evidence chains faster than it used to be. But the researcher still needs to verify that the links are accurate and that the quoted moments actually support the claim being made.

A simple policy that captures all of this: the tool can draft, the researcher decides. The AI is your assistant, not your replacement. If you are not reviewing the work, you are not doing research. You are forwarding emails.

The Directional Readout

The micro research deliverable is a directional readout. Not a report, not a deck, not a research document with a methodology section. A readout. Short, direct, structured, and designed to be consumed in minutes, not hours.

The readout answers four questions: What did you ask? What did you find? What does it mean for the decision? What can it not tell you?

The format varies by organizational culture. For some teams, it is a structured Slack message or a short document. For others, it is a one-page summary attached to a product brief. For still others, it is a verbal readout in a stand-up, with a written follow-up. The format matters less than the content. What matters is that the readout contains the findings, the supporting evidence, the boundaries, and the expiry. If it has those four things, the format is a delivery mechanism, not a quality signal.

Findings should be stated as observations, not recommendations. "Seven of twelve participants did not understand that the toggle switched between annual and monthly pricing" is an observation. "You should redesign the toggle" is a recommendation. The observation is what the research produced. The recommendation is a judgment that combines the research with product context, business priorities, and engineering constraints that the researcher may or may not have. State findings as observations; let the decision maker do their job.

Boundaries are the most important section—and the one most likely to be skipped. This study tested comprehension with twelve participants who match the target audience. It did not test adoption intent, conversion likelihood, or long-term retention. This section is what prevents micro research from being misused. Without it, a twelve-person comprehension test becomes "users validated the feature" in the retelling, which is a different and much larger claim than anything the study actually supported.

The expiry condition should be explicit. These findings are valid until the screen changes, the pricing changes, the flow is redesigned, or a specified amount of time passes—whichever comes first. Micro research has a short shelf life by design. Stating the expiry prevents findings from being cited in planning documents months later as if they are still current.

The entire readout should be writable in under an hour. If it is taking longer, either the study was not scoped tightly enough or you are overproducing the deliverable. Micro research earns its value by being fast to run and fast to deliver. A readout that arrives three days after the data came back has already lost most of its value. The decision it was meant to inform has already been made.

Microtools and the Execution Layer

The twenty-four- to seventy-two-hour window that defines micro research is achievable—but it requires something the operating model chapters do not address directly: an execution layer that eliminates the friction between raw data and usable input.

I introduced the concept of microtools on my blog,[11] and the response from practitioners confirmed what I had observed in my own work: the bottleneck in fast research is rarely the research itself. It is the unglamorous middle layer: the participant data that lives in a separate file from the responses; the behavioral telemetry exports that use different identifiers than the research tracker the open-ended responses that need cleaning before coding can begin; the screenshots that need extracting and labeling before analysis is possible. None of these are hard problems. They are time problems. And in a twenty-four- to seventy-two-hour window, time is the only resource that cannot be recovered.

A microtool is a small, disposable script that does one job, runs a handful of times, and gets archived or deleted. It is not a platform. It is not infrastructure. It is a wrench built for one bolt. The concept emerged from the practice of AI-assisted coding (sometimes called vibecoding), where describing what you need to a coding assistant, getting working code back, and running it—without necessarily understanding every line—has become a viable workflow for non-engineers. For researchers, this unlocks a category of execution problems that previously had no good solution: too specific for any platform to have anticipated, too time-consuming to do manually at speed, and too temporary to justify building anything real.

11 https://www.thevoiceofuser.com/some-thoughts-on-vibecoding-for-uxr-improving-project-execution-workflow-with-disposable-microtools/

What makes microtools useful in micro research, which is either a happy naming accident or a sign that I need to get out more, is that they protect researcher time for the work that cannot be automated. Data preparation, behavioral data integration, media extraction and labeling: these are tasks that consume hours when done manually and minutes when handled by a purpose-built script. The synthesis, the interpretation, the judgment calls about what the findings mean and what they cannot support, those remain the researcher's job. Microtools clear the path to that job.

The discipline that makes this work is narrow scope. One problem, one input, one output. A script that attempts to be a general solution collapses before it ships. The question before building anything is simple: what is the single thing this needs to do? If the answer takes more than two sentences, the scope is too broad.

Two constraints are nonnegotiable. The first is integrity checks: counts before and after transformations, totals that should be preserved, a random sample verified by hand. Research data processed by code you wrote in twenty minutes deserves the same skepticism you would apply to any other analytical step. The second is privacy as a hard limit: raw participant data stays local, PII does not go into prompts, and synthetic examples are used when generating code for anything that touches sensitive material. The speed that microtools provide does not change the ethics of how participant data is handled.

The cumulative effect is not just faster execution. When data preparation takes minutes instead of hours, the threshold for what is worth investigating drops. The extra segmentation cut gets run. The second data source gets pulled in to triangulate. The question that previously was not worth the setup cost gets explored. Over a quarter, the decisions being made are supported by richer evidence than they would have been without the execution layer working cleanly underneath. For micro research, that is the difference between a mode that hits its timeline and one that only aspires to it.

Common Pitfalls in Micro Research

Micro research has its own failure modes, distinct from sprint research. Knowing them does not prevent them entirely, but it helps you recognize them before they ruin a study.

The most common pitfall is scope ambiguity disguised as a tight question. A stakeholder asks, "Do users like the new dashboard?" That sounds specific. It is not. "Like" is not a measurable construct in a five-minute AI-moderated session. Do they understand it? Can they find what they need? Do they prefer it to the old version? Each of those is a micro research question. "Do they like it" is three questions in a trench coat pretending to be one. The question hygiene step exists precisely to catch this. If you cannot define what a useful answer would look like before the study runs, the question is not ready.

A second pitfall is skipping the first-three-participant audit. The study launches; all fifteen participants respond within a day, and the researcher goes straight to synthesis. Only then do they discover that a prompt was misunderstood, a stimulus was unclear, or the probes produced thin data. Fifteen responses, all compromised. The audit after three participants is a small investment that prevents this. Use it.

A third pitfall is treating the AI-generated summary as the finding. The summary says participants understood the pricing page. The researcher ships that conclusion in the readout. But the summary smoothed away two participants who were confused, because their confusion was a minority pattern and the AI optimized for the majority signal. Those two participants might represent a real segment with a real problem—or they might be noise. The researcher's job is to check and decide. The summary's job is to save time, not to replace judgment.

A fourth pitfall is overclaiming from directional data. Twelve participants understood the screen. The readout says "users understand the screen." That is a generalization the data cannot support. Twelve participants who match the target audience understood the screen in a controlled research context. Whether all users will understand it in the wild—on different devices, under different conditions, with different levels of attention—is a different question. The boundaries section of the readout exists to prevent this inflation, but it only works if the researcher actually writes it and the stakeholder actually reads it.

A fifth pitfall—and this is organizational rather than methodological—is letting micro research become the default for everything. When micro is fast and easy, every question starts to look micro-shaped. Deep research questions get crammed into twenty-four-hour studies; sprint -level questions get compressed into tight prompts. The taxonomy from Chapter 4 exists to prevent this. If you find yourself running micro studies that keep surfacing more questions than they answer, that is a signal. The question was not micro. It needed more room. Route it correctly next time.

Where This Goes Next

Micro research is the fastest mode. It is the most constrained. It works because the scope is tight, the question is specific, the quality gates are in place, and the readout is honest about what the findings can and cannot support. It is not a replacement for sprint research or deep research. It is a complement to them: the tool you use for the narrow, near-term questions that would otherwise go unanswered.

With this chapter and the previous one, the two fast operating modes are defined: sprint research for medium-complexity questions over one to two weeks, and micro research for narrow questions in twenty-four to seventy-two hours. Both are AI-assisted. Both require researcher judgment. Both produce findings with explicit boundaries and expiry dates. The difference is scope, depth, and the degree of researcher involvement in data collection.

The remaining chapters shift from operating modes to operating systems. The next chapter is where the two modes meet organizational reality. How do you route questions in real time when the request is ambiguous and the stakeholder is impatient? How do you handle the VP who wants a "quick study" on something that is clearly not quick? How do you say no without burning a relationship? How do you teach stakeholders to self-triage so that not every question requires negotiation? Chapter 8 is the practical chapter: using the two modes inside an actual organization with real politics, real ambiguity, and real pressure to treat everything as urgent.

CHAPTER 8

Using the Research Engine in Practice

Chapters 6 and 7 defined the two fast operating modes individually. Each chapter walked through the full arc: scoping, design, execution, analysis, delivery. But running micro and sprint research in practice involves skills that cut across both modes, along with situations that neither chapter addressed in isolation. How do you route a request when it arrives as a side comment in a planning meeting? How do you sample for directional work without pretending the sample is something it is not? How do you work with AI moderation tools without losing control of the data? How do you synthesize without overclaiming? How do you say no to a request that does not fit either mode?

This chapter is the practitioner's companion to the two mode chapters. It covers the shared craft layer that makes both modes credible and the organizational skills that make them sustainable. Chapters 6 and 7 explained what to do; this chapter explains how to do it well, repeatedly, in an environment that does not always cooperate.

The Real-Time Routing Conversation

Chapter 4 gave you the taxonomy and the routing logic: clean, systematic, logical. If research requests arrived in neat packages with labels attached, Chapter 4 would be the only routing guidance you need.

They do not arrive in neat packages. They arrive as Slack messages that start with "quick question"; as a PM sliding into your DMs with a screenshot and a deadline; or as a VP mentioning in a review that "we should really understand how users feel about this," followed by three people turning to look at you. Routing in practice is part classification, part negotiation, and part education—often happening in real time, without the luxury of a formal intake process.

The skill is pattern recognition. After enough practice with the taxonomy, you start to hear the routing dimensions in the way people describe their questions. "We just need to know if users get it" signals low ambiguity, probably low risk, and short expiry; that sounds like micro. "We are not sure what the right approach is" signals high ambiguity; that is deep research, regardless of what timeline the stakeholder has in mind. "We have a few options and we need to understand which one works better and why" suggests moderate ambiguity and moderate risk—sprint territory.

What makes this hard is that stakeholders do not describe their questions in terms of risk, ambiguity, and expiry. They describe them in terms of what they want and when they want it. "Can you test this by Friday" tells you the desired timeline but nothing about the actual routing dimensions. Your job is to listen past the timeline request and hear the question underneath: What are they actually trying to learn? How much is at stake? How well does the team understand this space? Those are the routing inputs. The timeline is a constraint, not a classification.

A practical approach: when a routing conversation starts, ask three questions internally before responding. First, what decision would this inform? If there is no decision, the study probably should not run. Second, what happens if they get it wrong? That gives you the risk level. Third, how well does the team already understand this space? That gives you the ambiguity level. With those three answers, you can usually route in under a minute.

How you communicate the routing matters. "I can run a micro study on this by Thursday that will tell you whether users understand the screen. It will not tell you whether they will adopt the feature or whether this is the right approach overall. Is that useful?" That response confirms you will help, sets a timeline, names what the study will deliver, and clarifies what it will not. The stakeholder can accept, push back, or refine. All three are productive outcomes. Compare that to "sure, I will look into it," which creates no shared understanding and almost guarantees disappointment when what you deliver does not match what they imagined.

When the Request Does Not Fit Either Mode

Not every request is a research question. Some are analytics questions. Some are design opinions looking for validation. Some are political maneuvers dressed up as curiosity. Part of routing is recognizing the requests that do not belong in the research pipeline at all and redirecting them cleanly.

Analytics questions are the most common. "Why did conversion drop last week?" is not a question research can answer. It is a question for the data team. But it lands on your desk because you are perceived as the person who "understands users," and a metric drop feels user-related. The redirect is straightforward: "That sounds like a measurement question. The data team can tell you what changed in the funnel. If the data suggests a user experience issue and you want to understand why users are behaving differently, that is where I come in." You are sequencing, not refusing—data first, then research if needed. Most of the time, the data answers the question and the research request evaporates.

Validation requests are subtler. The decision has already been made. The stakeholder wants research to confirm it. The tell is in the language: "We just need to validate that users are okay with this." Validation is not inherently illegitimate. Sometimes a team has high confidence and wants a lightweight check before shipping. That can be a fine micro study. The problem arises when the team has no intention of changing course regardless of what the research shows. At that point, the study becomes theater.

The way I handle this is by asking the outcomes question: "If the study shows that users do not understand this, what will the team do?" If the answer is "we will revise it," the study is real. If the answer is hesitation, I reframe: "It sounds like the decision is made. Instead of a validation study, would it be more useful to identify the specific points of confusion so the team can address them in a follow-up iteration?" The data collection stays the same, but the framing—and therefore the utility—changes. A useless validation exercise becomes a useful comprehension study.

Political requests are the hardest to handle because they are rarely explicit. Nobody says, "I need research to prove my colleague wrong." Instead, they say, "I think it would be really valuable to understand how users experience this flow," where the flow happens to be the thing their colleague championed. When you can detect this, the best approach is to scope the study so that it produces useful evidence regardless of which internal position it supports. If the findings are genuinely useful for the product, it does not matter that the initial motivation was political.

Escalation Triggers

The taxonomy is not a one-time classification. It is a continuous assessment. A question that looks like micro at intake can reveal itself as something larger once data starts coming in. Knowing when to escalate is one of the most important practical skills in this work.

The most common escalation is micro to sprint, and the signals are recognizable. Participants give long, complex answers to what should be simple questions, suggesting the ambiguity was higher than scoping assumed. Participants contradict one another in ways that cannot be explained by normal variation, suggesting structural differences across user types that require follow-up conversations. The first-three-participant audit reveals that the prompts are producing data, but the data is raising more questions than it answers.

When these signals appear, you have three options: finish the micro study and deliver what it can deliver, with an explicit note that the findings suggest a deeper question warranting a sprint study; stop the study and redesign it as a sprint, which makes sense when the micro data is compromised enough that the directional signal is unreliable; or finish and immediately scope a follow-up sprint while the data is fresh. The decision contract is what makes this manageable. Because you defined what the study would and would not conclude before it started, you have a clear basis for saying, "This is beyond what we scoped."

The escalation that matters most—and happens least—is escalation to deep research. When micro or sprint studies repeatedly surface the same confusion, the same unexpected mental models, or the same disconnect between what the team assumes and what users actually think, that is a signal that the Frame is wrong. The team is asking specific questions inside a Frame that does not match reality. No amount of micro or sprint research fixes a broken frame. Only deep research does. A broken Frame means the team is optimizing the wrong thing with increasing precision. Flagging this is uncomfortable because it means telling the team its foundational understanding needs work—but not saying it means continuing to produce precise answers to the wrong questions.

Sampling for Directional Work

Every research method has a sampling logic. Surveys have statistical sampling; ethnography uses theoretical sampling. Micro and sprint research rely on what I call directional sampling: recruiting enough participants with enough variation to see a pattern and bound it, without making population-level claims from the result.

The audience definition comes first, and it needs to be tight: who are you trying to learn from, and why them? If the question is about onboarding, you need new users—not power users who have not seen the onboarding flow in two years. If the question is about pricing, you need people who match the buyer profile your product actually serves, not a convenience sample of whoever responds fastest. Loose audience definitions produce data that is technically from real users but practically from the wrong users. That is a subtle form of garbage in, garbage out.

Sample size should be proportionate to the mode. For micro research, eight to fifteen participants is typical; for sprint research, fifteen to twenty in the AI-moderated phase plus four to six follow-ups. These are not magic numbers. They are ranges that produce enough data to see patterns without investing disproportionate effort for the decision at hand. If you find yourself recruiting thirty people for a micro study, either the question is not micro or you are overinvesting.

The concept that is more important than raw count is minimum viable diversity. You do not need a representative sample. You need enough variation across the dimensions that matter for the question. If you are testing a pricing page, variation in business size, familiarity with the product category, and price sensitivity matters more than demographic balance. If you are testing a workflow, variation in technical confidence and prior experience with similar tools matters more than geography. The dimensions of diversity should be driven by the question, not by a generic demographic checklist.

A practical rule: identify the two or three dimensions most likely to produce different experiences, and make sure your sample varies along those dimensions. You do not need equal representation. You need enough variation to see whether the pattern holds or breaks across the dimensions that matter.

What you cannot do—and should never pretend to do—is generalize from these samples to populations. The language in your readout should be precise: "among the participants in this study" or "the participants who matched our target criteria," not "users think" or "users prefer." The definite article makes a universal claim. The bounded phrasing makes a directional one. That difference matters, and it is the researcher's job to enforce it—even when stakeholders would prefer the more confident version.

Recruitment Source and Participant Quality

Where your participants come from affects data quality more than most researchers acknowledge in their readouts.

The best source is your own users, recruited directly from your product. They have real context, real usage history, and real stakes. When they tell you the pricing page is confusing, they are reacting to pricing they might actually pay; when they say the onboarding makes sense, they have just onboarded for real. This gives the data ecological validity that panel-sourced data cannot match.

The second-best source is targeted recruitment from your user base or customer list, where you screen for the specific characteristics the study needs. This requires more coordination but yields participants who match the audience definition precisely.

The third source, and the one most studies end up using because of speed constraints, is a research panel. Panels provide fast access to participants who match demographic and behavioral criteria, but the trade-off is that many panel participants are professional research participants. They have learned what researchers want to hear. They are optimized for completing studies quickly, not for giving you the messy, honest, context-rich responses that make qualitative data useful. Some panel participants are excellent. Some are going through the motions. The quality variance is higher than with your own users.

Then there is the fraud problem. Panel recruitment, especially at speed, attracts participants who misrepresent themselves, people who lie on screeners to qualify for more studies, people who use AI to generate responses, or people who complete studies in a fraction of the time it should take because they are not actually reading or thinking. Tools are building detection mechanisms, but the researcher still owns the quality gate. If a participant's data looks suspect, exclude it before synthesis. Do not let bad data contaminate findings simply because removing it feels like you are shrinking your already small sample.

A practical standard: state the recruitment source in every read-out. "Participants were recruited from our active user base" carries different weight than "participants were recruited from a general research panel." Both are legitimate; neither should pretend to be the other. The decision maker deserves to know where the data came from.

Working with AI Moderation Tools

AI-moderated data collection is the engine that powers both micro and sprint research. Tools such as Outset handle session logistics—presenting prompts; asking follow-up probes; capturing responses in text, audio, or video; generating transcripts with timestamps; and producing initial summaries. The researcher designs the study and interprets the results; the tool executes the middle.

This division of labor is powerful but requires understanding what the tool controls and what you control, and where the handoff points create risk.

You control the prompt design; the tool executes the prompts as written. If your prompts are vague, the tool will ask vague follow-ups and produce vague data. If your prompts are specific and grounded in concrete artifacts, the tool will produce specific, comparable data across participants. The quality of the AI-moderated data is largely a function of prompt quality. The tool scales your questions; it does not fix bad questions.

You control the guardrails. AI moderation tools allow you to set boundaries on how the AI probes, what topics it avoids, how it handles unexpected responses, and when it moves on. These guardrails matter because without them, the AI will sometimes follow tangents that participants introduce, ask leading follow-ups that shape responses, or probe into areas that are outside the study scope. Setting guardrails is not optional. It is part of the study design, and it should happen before launch, not as a reaction to bad data.

The tool controls the moderation flow. It decides, based on your prompts and guardrails, how to probe individual responses. This is where the risk of leading or shaping comes in. AI moderators can inadvertently suggest answers, validate participant responses in ways that discourage honesty, or push for elaboration in ways that feel pressured. The mitigation is the first-three-participant audit: review the first three responses in full, including the AI's probes, and check whether the moderation is neutral or shaping. If the AI is leading, adjust the guardrails before the rest of the data comes in.

The tool also generates summaries. These summaries are useful as a starting point for synthesis. They are not findings. They are drafts.

Summaries smooth away contradiction, inflate agreement, and optimize for clean narratives. That is what summarization algorithms do. Your job is to read past the summary to the raw data, check whether the summary accurately represents what participants actually said, and catch the moments where the summary missed something important because it did not fit the majority pattern.

One practical note on tool selection: the specific tools will change. Outset exists now and works well for what it does, but others will emerge. Some will be better and some will be worse. The principles in this chapter are tool-neutral. Whatever tool you use, the same questions apply. Does it give you full transcripts with timestamps? Does it let you set moderation guardrails? Does it link summary claims to specific participant moments? Does it let you audit the AI's probing behavior? If yes, the tool is usable. If it produces summaries without traceable evidence, it is not usable for serious research regardless of how polished the interface is.

Evidence Standards Across Both Modes

Whether you are running a micro study or a sprint study, the evidence standards are the same. The difference is the volume of data and the time available for synthesis, not the rigor of the claims you make.

Evidence chains are the foundation. Every claim in a readout should be traceable to specific participant moments: a quote, a clip, a timestamp. This is not academic formalism. It is practical protection. When a stakeholder questions a finding, you need to produce the supporting evidence in under a minute. When a finding gets cited in a road map discussion three months later, someone needs to be able to verify what it was actually based on. Without evidence chains, findings become folklore—things the organization believes because someone said them in a meeting, with no way to check whether they were ever true.

Transcript auditing is the quality check that makes evidence chains trustworthy. The AI generates a summary; the summary says something; you check the transcripts to see whether the summary accurately reflects what participants said. This sounds tedious, but it is not optional. AI summaries have specific failure modes. They smooth away contradiction, treating two opposing views as minor variation rather than genuine disagreement. They inflate agreement, making a finding sound more universal than it is. They miss quiet signals because a participant who mentions something briefly does not register as strongly as a participant who talks about it at length. Auditing catches these failures before they become findings.

Disconfirming evidence deserves specific attention because it is the quality marker that separates research from storytelling. Every theme in your synthesis should have at least one disconfirming example. If your finding is that users understood the pricing page, find the participant who did not. If every participant understood the page, that is a genuinely strong finding, and you can say so. But the absence of disconfirming evidence should be verified, not assumed. Look for it. If it is not there, good. If it is there and the AI summary missed it, you just caught the most important quality failure in the entire workflow.

Observation versus inference is the cleanest split you can make in synthesis, and it is the one most researchers blur without realizing it. An observation is what the participant did or said. "Seven of twelve participants could not find the monthly pricing toggle" is an observation. An inference is what you think it means. "The pricing page design obscures the monthly option" is an inference. Both may be correct, but they are different types of claims, and they should be labeled differently in the readout. Observations rest on evidence; inferences rest on judgment. Mixing them without distinction is how directional findings get treated as established facts.

Frequency versus importance is another distinction that deserves conscious tracking. Five participants mentioning a minor annoyance does not automatically make it the most important finding. One participant describing a workflow-breaking problem that nobody else encountered might matter more for the decision at hand. Frequency tells you how common something is; importance tells you how much it matters. Micro and sprint research are small-sample methods. Counting heads and ranking by frequency misuses the data. Weigh what you heard by its relevance to the decision, not by how many times you heard it.

Communicating Directional Findings

The deliverable for both micro and sprint research is a bounded, directional communication. The format differs, as Chapter 6 and 7 described, but the communication discipline is the same: state what you found, what it means for the decision, what it cannot tell you, and when it expires.

Precision language matters more in directional research than in any other type. Because the sample is small and the scope is tight, every word in the readout either maintains the appropriate level of confidence or inflates it. "Users prefer option A" is inflation. "Most participants in this study chose option A, primarily because of the clearer pricing display" is precision. The first version will be quoted in a product review as "research says users prefer A." The second version is harder to misquote because the qualifications are built into the sentence.

Expiry dates are not a formality. They are a functional mechanism for preventing misuse. A micro study on how users experience a pricing page is valid until the pricing page changes. A sprint study on onboarding friction is valid until the onboarding flow is redesigned. State the expiry condition in the readout: "These findings are valid for the current version of the pricing page. If the page is redesigned, the findings no longer apply." This sounds obvious, but in practice, findings from six months ago get cited in current road map discussions constantly. The expiry condition is your defense against that.

Mixed evidence requires careful presentation. If eight participants understood the flow and four did not, you do not have a clean finding—you have a split. Presenting this as "most users understood" obscures the four who did not. Presenting it as "results were mixed" obscures the eight who did. The honest presentation is: "Eight of twelve participants completed the flow without difficulty. Four encountered confusion at the toggle step, primarily because they did not realize it controlled the pricing display." That gives the decision maker both the pattern and the exception, which is what they need to decide whether to ship, revise, or investigate further.

One communication pattern I have found effective across both modes: lead with the finding, follow with the evidence, and close with the boundary. "Participants did not understand that the toggle switched between monthly and annual pricing (seven of twelve). The most common interpretation was that the toggle controlled a different plan tier entirely (quotes: P3, P7, P11). This finding applies to the current design and was tested with users who had not previously seen the pricing page." That is three sentences. It contains a finding, evidence, and a boundary. It takes thirty seconds to read and sixty seconds to verify. That is the standard.

When Findings Contradict Across Studies

This will happen. You run a micro study that says users understand the pricing page. Two weeks later, a sprint study surfaces deep confusion about the pricing model. Both studies were well-designed. Both produced credible findings. A stakeholder asks which one is right.

Usually, both are right. The contradiction is not a quality failure. It is a scope difference. The micro study tested comprehension of a specific screen in a controlled context. The sprint study explored how users think about pricing in the context of their actual purchasing decision. Users can understand what the page says and still be confused about what the pricing means for them. These are different findings about different things, and they only contradict each other if you treat them as answering the same question.

The first step is to check whether the studies actually addressed the same question. Most of the time, they did not. The micro study answered a narrow question. The sprint study answered a broader one. The apparent contradiction dissolves when you restate each finding within its actual scope.

When the studies genuinely do address the same question and produce different results, the most common explanation is that the sprint study surfaced something the micro study was not designed to capture. The micro study's scope was too tight to see it. That is not a failure of the micro study. It is exactly the limitation that the boundaries section was designed to flag. The sprint finding supersedes the micro finding because it rests on more evidence with greater depth.

The communication to stakeholders matters here. Do not Frame it as "the first study was wrong." Frame it as "the first study answered the question it was designed to answer. The second study revealed a broader pattern that changes the picture." That framing maintains trust in both studies and in the research function. If stakeholders conclude that micro research is unreliable because a sprint study found something different, they will stop trusting directional findings entirely. The issue was never reliability. It was scope. Make that distinction explicit every time.

Saying No and Saying Not Yet

Routing is not just about matching questions to modes. It is also about recognizing when a question should not be studied at all—or not right now—and communicating that without losing trust.

The first principle: never just say no. A bare refusal with no alternative leaves the stakeholder with the same need and no path forward. They will either go around you, do the research themselves badly, or decide that research is unhelpful. All three outcomes are worse than the request that prompted the refusal.

Every no should come with one of four things. A reframe changes the question into something answerable: "I cannot tell you whether users like the dashboard, but I can tell you whether they understand it and where they get confused." A redirect sends the question to the right place: "That is a measurement question. The data team can answer it faster and better than I can." A smaller yes offers a reduced version: "I cannot do a sprint study in this timeline, but I can run a micro study on the comprehension piece and have it by Thursday." A future yes defers without dismissing: "This needs sprint-level depth and I cannot start until the fourteenth. In the meantime, here is what I would suggest the team does with what you already know."

The second principle: use the system, not personal judgment, as the basis for the refusal. "I do not think this needs a study" is a personal opinion that invites argument. "The question has low ambiguity and the stakes are low, so a micro study is proportionate. A sprint study would overinvest" is a system-level assessment. The taxonomy and the MVR framework give you language that depersonalizes the decision. You are not being difficult. The system is telling you what the question requires.

The third principle: protect the no after you say it. The most common failure is caving to pushback. The stakeholder says "but this is really important" or "the VP wants it." If the routing was correct, neither of those changes the assessment. Importance does not change the ambiguity level. Seniority does not reduce the rigor threshold. The risk of the decision determines the rigor required. If the timeline does not allow the required rigor, the answer is not to do insufficient research. The answer is to flag the gap and let the decision owner choose.

The VP's Quick Question

This gets its own section because it is a specific and recurring pattern, and the standard advice of "just push back" ignores the power dynamics involved.

The scenario: a senior leader asks a research question with an implicit expectation of speed. "Can we get some quick user input on this? I want to make sure we are not missing something before we commit." The question may be reasonable. The problem is the context: it comes from someone whose requests carry organizational weight, and declining can feel politically risky.

The first thing to recognize is that most senior leaders are not asking for bad research. They are asking for information to make a better decision. They default to "quick" because they assume research is slow and they are trying to be considerate. The conversation is not adversarial. It is a scoping conversation with someone who has less methodological context than you do.

A practical response: "That is a great question. I can run a micro study that tells you whether users understand the new pricing structure, and I can have directional findings by Thursday. What it will not tell you is whether the pricing is competitive or whether users will convert at the new price points. Those are bigger questions that would need a different approach. Is the comprehension piece what you need right now?"

That response shows you are responsive, not resistant. It offers a concrete deliverable with a timeline. It names the boundaries honestly. It ends with a question that puts the decision back with the leader. Most of the time, the answer is "yes, that is what I need." Occasionally, it opens a conversation about deeper work. Either way, you have routed the question professionally rather than just accepting or refusing it.

Teaching Stakeholders to Self-Triage

The long-term solution to routing overhead is not a better intake form. It is stakeholders who understand the taxonomy well enough to bring you better questions from the start.

This is not about training stakeholders to do research. It is about training them to think about their questions before handing them off. The same way a well-trained PM writes a clear brief before engaging engineering, a well-trained stakeholder should be able to articulate what they need to learn, what decision it connects to, and what kind of answer they need.

The most effective training happens through repeated interaction, not through workshops or documentation. Every routing conversation is a teaching moment. When you respond by saying "that sounds like a micro study because the question is specific and the stakes are low; here is what I can deliver by Thursday," you are teaching the taxonomy. When you say "that question is bigger than it looks; here is why it needs a sprint rather than a quick test," you are teaching ambiguity assessment. When you say "this is really a data question," you are teaching the boundary between research and measurement.

Over time, this produces stakeholders who self-triage. They start saying "I think this is a micro study" before you assess it. They start including decision owners and boundaries in their requests without being asked. They stop asking for quick studies on deep research questions because they have learned what those questions look like.

A practical tool that accelerates this: when you deliver a readout, include a one-sentence note on why the study was scoped the way it was. "This was run as a micro study because the question was specific, the risk was low, and the team needed an answer before the sprint ended." Those notes take ten seconds to write, and they reinforce the taxonomy every time a stakeholder reads a deliverable. They normalize the idea that different questions get different treatments, which is the foundational concept behind the entire system.

Where This Goes Next

This chapter covered the shared craft and the organizational skills that make micro and sprint research work in practice: routing Routing in conversation; sampling for directional work; working with AI tools without losing control; evidence standards; synthesis discipline; communication patterns; saying no; handling senior leaders; and training stakeholders over time.

All of this assumes a manageable volume of requests. But one of the consequences of making research faster is that demand for research increases. Stakeholders who never asked for research before start asking. Questions that were never considered researchable become "can you just run a quick study" requests. The calendar fills. The researcher becomes a bottleneck again—not because the work is slow but because there is too much of it.

That is the demand explosion, and it is the subject of the next chapter.

CHAPTER 9

Demand, Governance, and Quality at Scale

Here is the part nobody says out loud when they pitch AI-assisted research tools. The pitch is "do more with less." The reality is "do more with the same—and also explain why you cannot do even more."

This is not a hypothetical. It is a predictable consequence of making research faster and cheaper. When research took six weeks, stakeholders self-rationed. They only asked for research when the question was big enough to justify the wait. Most small questions went unanswered or got resolved by opinion. That was a problem, and a significant portion of this book is about solving it. But it was also, accidentally, a demand management system. The slowness of traditional research functioned as a filter. Only the most important questions made it through.

Remove the filter and everything comes through. Micro research takes two days. Sprint research takes two weeks. The cost to the stakeholder, in terms of time and attention, drops dramatically, so they ask for more. Questions that were never considered researchable become requests. PMs who never engaged with research start expecting one- to two-day turnaround on everything. Designers who used to rely on intuition start wanting validation for every screen. The research function goes from underutilized to overwhelmed—not because the work got harder but because the perceived cost of requesting it dropped to near zero.

This chapter covers both sides of the problem. The first half is about managing demand: intake, prioritization, service levels, and capacity. The second half is about governance: what happens when micro research is no longer just you running studies but something the organization relies on; when other people want to run studies; and when quality standards need to be organizational policy rather than personal habit. The two halves are connected. Demand creates pressure. Governance is the system that handles pressure at scale without killing quality.

The Economics of Perceived Cheapness

The demand explosion follows basic economics. When the perceived cost of something drops, demand increases. Research used to be perceived as expensive: slow, resource-intensive, requiring weeks of lead time. That perception suppressed demand. Many questions that would have benefited from user evidence were never asked because the cost of getting an answer seemed disproportionate to the decision.

AI-assisted research changes the perceived cost. A micro study takes two days. A sprint study takes two weeks. The researcher is not locked up in sessions all week. The tools handle the moderation.

The results come with summaries, timestamps, and linked quotes. It looks fast, and it looks easy. From the stakeholder's perspective, the constraint that used to prevent them from asking has been removed.

The problem is that the perceived cost dropped faster than the actual cost. A micro study takes two days of calendar time, but it still requires researcher time: question hygiene, prompt design, quality auditing, synthesis, the readout. A sprint study compresses data collection, but the researcher is still involved throughout: designing the AI-moderated phase, selecting follow-up participants, conducting follow-up sessions, integrating two data layers, producing a deliverable. The tools compressed execution. They did not compress judgment—and judgment is the bottleneck that stakeholders do not see.

The result is a gap between what stakeholders expect and what the researcher can deliver. The stakeholder sees a tool that can run a study in two days and concludes that the researcher can run a study every two days. The researcher knows that running a study every two days means doing nothing else: no synthesis depth, no quality auditing, no thinking time, no deep research, no strategic work—just a production line of micro studies, each technically complete and none particularly good.

This gap is where most of the tension in adopting AI-assisted research comes from. It is not a tools problem. It is a demand management problem, and solving it requires infrastructure, not just effort.

Intake That Filters Before It Assigns

The first piece of infrastructure is an intake system that prevents bad requests from consuming research time. Chapter 8 described a lightweight intake that captures the routing inputs: the question, the decision, the decision owner, and the timeline. That intake also serves as the first demand filter, because a surprising number of requests cannot survive those four fields.

A request with no decision is not a research request. It is curiosity. Curiosity is fine. But it does not get a study. A request with no decision owner is not actionable. Nobody will act on the findings, so it does not get a study. A request where the decision has already been made is validation theater. It gets reframed or declined. A request where the question is too vague to route gets sent back for scoping before it enters the pipeline.

The intake does not need to be heavy. It can be a Slack form, a shared document, or a standing question in planning. What matters is that every request passes through it and that the researcher sees the routing inputs before committing. The filtering happens naturally because most poorly scoped requests reveal themselves when the stakeholder has to answer four specific questions. If they cannot name the decision, the decision owner, or the question, they are not ready for research. That is not a rejection. It is a sequencing conversation: come back when you can answer these four things, and I will help you.

The filtering rate will surprise you. In my experience, somewhere between one-third and one-half of incoming requests are not ready for research when they arrive. Some need scoping help. Some are analytics questions. Some are already answered by existing data.

Some are not connected to any real decision. Catching them at intake saves an enormous amount of time compared to catching them mid-study—or worse, delivering findings that nobody uses.

Prioritization When Everything Is Urgent

Intake filters the requests that should not be studies. Prioritization addresses a harder problem: when you have more legitimate requests than capacity to run them.

The temptation is to prioritize by urgency. The request with the tightest deadline goes first. This feels responsive and collaborative. It is also a trap, because urgency is a feature of the stakeholder's timeline, not a feature of the question's importance. A low-risk comprehension test that someone needs by Friday is urgent but not important. A sprint study on a new pricing model that will affect revenue for the next year is important but might not feel urgent because the launch is two months out. If you prioritize by urgency alone, the important work gets perpetually deferred by the urgent work.

A better prioritization logic uses the same dimensions as routing but applies them to the queue rather than to individual studies. Risk first: which decisions carry the highest consequences if the team gets them wrong? Those get priority regardless of urgency. Then alignment: which studies connect to the organization's current priorities, OKRs, or strategic bets? Research exists to support decision-making, and the decisions that matter most to the organization should get research first. Then urgency: among studies with similar risk and alignment, the one with the tighter timeline goes first.

This is not a scoring formula. It is a thinking tool. The point is to have a principled basis for saying "I am doing this study first because the decision it informs has higher stakes and stronger alignment to our current goals," rather than "I am doing this study first because it was requested most recently." The first is a professional assessment. The second is a queue.

One practical implication: prioritization requires visibility into the full pipeline. If you are routing and prioritizing one request at a time, you are making local decisions without seeing the global picture. A simple running list of current and pending studies, with the routing assessment for each, gives you that visibility to make trade-offs. When a new request arrives, you can see what it displaces. When a stakeholder asks why their study has not started, you can show them what is ahead of it and why. Transparency is what makes prioritization decisions defensible.

Service Levels and the Calendar

One of the most effective demand management tools is also one of the simplest: published service levels that tell stakeholders what to expect before they ask.

A service level is a commitment about turnaround time for each research mode. Micro research delivers directional findings within two to three business days of a scoped request. Sprint research delivers findings within one to two weeks of kickoff. Deep research has a timeline scoped per project, typically four to eight weeks. These are not guarantees. They are norms. They tell stakeholders what "fast" actually means, which prevents the magical thinking that produces requests like "can I have a sprint study by Friday."

Published service levels do two things. First, they set expectations. A stakeholder who knows that micro research takes two to three days will scope their requests accordingly. They will not ask for a micro study the afternoon before a decision meeting, or if they do, they know they are asking for an exception, not a normal delivery. Second, they create accountability in both directions. The researcher commits to delivering within the service level. The stakeholder commits to providing the routing inputs with enough lead time for the service level to work.

Service levels also make prioritization conversations easier. When you have to defer a study, you can point to the service level: "Micro studies have a two- to three-day turnaround. I have two studies ahead of yours. I can start yours on Wednesday and have findings by Friday. Would that work, or do we need to reprioritize?" That conversation is concrete and productive. Without service levels, the same conversation is vague and frustrating for both sides.

One nuance: service levels should reflect what you can actually sustain, not what you wish you could deliver. If you publish a two-day turnaround but regularly deliver in four because you are overloaded, the service level loses credibility and stakeholders route around you. Be honest about your capacity. If the honest service level is longer than stakeholders want, that is a useful data point for a capacity conversation with your leadership—not a reason to make promises you cannot keep.

The calendar is the enforcement mechanism for both service levels and prioritization. Your calendar is not a to-do list. It is a strategic resource. How you allocate your time across the three modes determines what kind of evidence the organization has access to. If you fill your calendar with micro studies, the organization gets a lot of directional signal on small questions but nothing deeper. If you protect time for sprint and deep research, the organization gets foundational understanding but might have gaps in near-term coverage.

A practical discipline: block time for non-micro work before the micro requests fill it. If a sprint study needs to happen in the next two weeks, block the follow-up session days and synthesis time now. Micro studies, because they are short and responsive, can fill around the blocks. But if you do not block first, the micro studies consume everything because they are always there and they are always "just two days." Look at your calendar for the last quarter. Count the micro studies, the sprint studies, the deep research projects. That ratio is your actual research strategy, regardless of what you said your strategy was. If the ratio does not match what the organization needs, the calendar is the thing to fix.

When Demand Exceeds Capacity

Even with intake filtering, prioritization, and service levels, there will be periods when demand exceeds capacity. When there are more legitimate requests than you can run. The question is not whether this will happen. It is what you do when it does.

The wrong answer is to work faster. Working faster in research means cutting corners: skipping the first-three-participant audit, shipping AI summaries without transcript review, compressing synthesis, dropping the disconfirming evidence check. You produce more studies, but each one is less trustworthy. Over time, stakeholders notice that the findings are thin or contradictory or get proven wrong. Trust erodes and the research function loses credibility—not because it was slow but because it was fast and wrong.

The wrong answer is also to work more hours. Sustained overwork degrades judgment, which is the one thing the tools cannot replace. A researcher running three micro studies simultaneously while starting a sprint study is not doing any of them well. The quality gates are the first casualty of overload because they are the steps that feel optional when you are behind.

The right answer is to make the overload visible and let the organization make the trade-off. "I have capacity for two micro studies and one sprint study this week. There are four micro requests and two sprint requests in the queue. Here is how I would prioritize based on risk and alignment. If the team wants a different order, we can discuss. If the team wants all of them done this week, we need to discuss capacity."

That conversation is uncomfortable. It is also the researcher's job. Making the trade-off invisible—by quietly cutting quality or burning yourself out—is worse for everyone. The organization does not learn that it is underinvesting in research capacity. The researcher does not get the support they need. The work gets worse without anyone understanding why.

The capacity conversation is where data helps. If you have been tracking your intake, you can show the trend: "Incoming requests have increased 40 percent since we adopted micro research. My capacity has not changed. Here is the volume, here is the current queue, here is the average turnaround. We are at the point where adding more requests means either longer turnaround or lower quality." That is a professional assessment backed by data, not a complaint about workload. It gives leadership the information they need to make a resourcing decision.

The Governance Question

Everything up to this point in the chapter has assumed that you are the one running the studies. Demand management is about how you handle the volume of requests coming to you. But as micro research becomes established and the tools become familiar, a different pressure emerges: other people want to run studies too.

PMs see how micro research works and think—reasonably—that they could do it themselves for simple questions. Designers want to test comprehension on their own screens without waiting for a researcher to be available. Research leads with multiple team members want to scale the practice without becoming a bottleneck. The question shifts from "how do I manage demand on my time" to "how does the organization manage research quality when multiple people are running studies."

This is the governance question, and it is a fundamentally different problem from demand management. Demand management is about capacity and prioritization. Governance is about quality and standards. You need both, and they interact. The demand pressure creates the incentive to distribute research to more people. The governance system determines whether that distribution produces useful evidence or organizational noise.

The worst outcome is ungoverned distribution: multiple people running micro studies with no shared quality standards, no intake process, and no evidence requirements. Findings are produced, presented, and acted upon without anyone auditing whether they are credible. The organization thinks it has more user evidence than before, when in reality, it has more research-shaped output, some of which is good and some of which is garbage, and nobody can tell the difference. That is worse than having no research at all, because at least with no research, the team knows it is guessing. With bad research, the team thinks it has evidence when it does not.

Quality Gates as Organizational Policy

In earlier chapters, quality gates were described as personal practice: things the researcher does to keep their own work credible. Transcript auditing; disconfirming evidence; evidence chains; boundaries language; expiry dates: when you are the only one running studies, these are habits. When multiple people are running studies, they need to become policy.

The distinction matters because habits are optional. A PM running a micro study for the first time does not have the habit of auditing AI-generated summaries against raw transcripts. They do not have the habit of looking for disconfirming evidence. They do not have the instinct that a finding without an evidence chain is an impression, not a conclusion. These are not character flaws. They are skills that researchers develop through training and practice, and expecting non-researchers to have them by default is unrealistic.

Organizational quality gates are the minimum standards that apply to any study, regardless of who runs it. They are not aspirational. They are nonnegotiable. If a study does not meet the quality gates, its findings are not treated as research evidence in decision-making.

The gates I recommend as organizational policy are deliberately simple, because complexity reduces compliance. First, every study must have a named decision and a named decision owner before it launches. If there is no decision, there is no study. Second, every finding in a readout must be linked to specific participant evidence: a quote, a clip, a timestamp. If a finding cannot be traced to a participant moment, it is an impression, not a finding. Third, every readout must include at least one disconfirming example per theme, or an explicit statement that no disconfirming evidence was found after looking for it. Fourth, every readout must include explicit boundaries on what the findings can and cannot support. Fifth, every readout must include an expiry condition.

Five gates. None of them require advanced research skills. All of them can be taught in an hour. And all of them, if followed, prevent the most common quality failures: studies with no purpose, findings with no evidence, conclusions with no boundaries, and research that gets cited long after it has expired.

The enforcement mechanism matters. If the quality gates are documented but nobody checks compliance, they are guidelines, not policy. The minimum viable enforcement is a review step: before a study's findings are shared broadly or used in a decision, someone with research training reviews the readout against the five gates. That review can be fast—fifteen minutes for a micro study readout—but it needs to happen. Without it, the gates are aspirational and compliance will erode within weeks.

The Self-Serve Question

Can PMs and designers run their own micro studies? The answer is yes—with conditions, and those conditions are not optional.

The case for self-serve is straightforward. The researcher is a bottleneck, and many micro research questions are simple:

comprehension tests, preference checks, terminology validation. The tools are designed to be usable by non-researchers, so if a PM can write a clear question, define a target audience, and launch a study, why should they wait for a researcher who has three other studies in the queue?

The case against self-serve is also straightforward. The researcher's value is not in operating the tool. It is in question hygiene, prompt design, quality auditing, and interpretation. A PM who can operate Outset can collect data, but whether that data is useful depends on whether the question was wellscoped, the prompts were well designed, the participants were well selected, and the synthesis was well audited. Those are research skills, and the tool does not provide them.

The resolution is conditional self-serve: non-researchers can run studies under specific conditions, with specific guardrails, and with a review step before findings are treated as evidence.

The conditions I recommend are these. The question must pass intake—named decision, named owner, clear question, defined timeline—which prevents studies that have no purpose. The study must use approved prompt templates or have its prompt set reviewed by a researcher before launch. This prevents poorly designed studies from producing misleading data. The readout must meet the five quality gates, and the readout must be reviewed by a researcher before the findings are shared broadly or used in a decision.

That last condition is the critical one, and the one that generates the most pushback. The whole point of self-serve is speed, so adding a researcher review step feels like reintroducing the bottleneck. But the review is not the bottleneck—it takes fifteen minutes. The bottleneck was the researcher designing, running, and synthesizing the entire study. The difference between "the researcher runs the study" and "the PM runs the study and the researcher reviews the readout" is significant: the first takes two to three days of researcher time; the second takes fifteen minutes. That is a massive capacity gain—it just requires that the fifteen-minute review actually happens.

The alternative to conditional self-serve is unconditional self-serve, and it is worth being blunt about what that produces. PMs run studies with poorly scoped questions; the AI-generated summaries become the findings; those findings are presented in meetings without evidence chains, without disconfirming examples, without boundaries. Stakeholders act on them. Some of the findings are correct. Some are not. Nobody can tell which is which because there is no quality standard. Over time, the organization develops a corpus of "research findings" of unknown quality, and the researcher spends more time correcting bad research than running good research. That is not a theoretical risk—it is the predictable outcome of distributing research without distributing quality standards.

A useful framing for the self-serve conversation: the tool is a power tool. Anyone can use a power tool, but not everyone should use one unsupervised. The training is quick; the safety standards are clear; the supervision is light—but it exists, because the consequences of using a power tool badly are worse than not using it at all.

Escalation Decision Trees

The preceding conditions define when a study can be self-served. The escalation decision tree is the mechanism that makes the distinction visible and fast for the person deciding whether to launch.

The simplest version maps directly to the routing dimensions from Chapter 4. For low risk, low ambiguity, and short expiry, it's self-serve eligible—with researcher review of the readout. For moderate risk or moderate ambiguity, the researcher reviews the study design before launch, not just the readout. For high risk, high ambiguity, or long expiry, it's researcher-owned. Period.

The tree should also include trigger conditions during a study. If the self-serve study reveals unexpected complexity, if participants are responding in ways that suggest the question is bigger than anticipated, or if the data contradicts the team's assumptions strongly enough that the implications are significant, the study should escalate to researcher involvement. The PM or designer running the study may not recognize these signals, which is why the researcher review step matters even for studies that seem straightforward at launch.

Make the tree visible and simple: a one-page document or a decision flowchart that anyone running a study can reference. It should produce a clear answer in under a minute. If it requires extensive judgment to navigate, it is too complicated, and non-researchers will ignore it.

Evidence Standards Across the Organization

When research was something only researchers did, evidence standards were implicit—the researcher knew what credible findings looked like because they were trained to produce them. When research becomes something multiple people do, those standards need to be explicit, shared, and enforced.

The quality gates described earlier in this chapter are the minimum bar, but evidence standards go beyond the gates. They shape how the organization thinks about and uses research evidence in general.

One standard that matters more than most organizations realize is the distinction between research evidence and other kinds of input. A micro study with twelve participants is research evidence. A PM's conversation with two customers at a conference is anecdotal input. A stakeholder's intuition from years of domain experience is expert judgment. All three are valuable, but none of them is the same thing. When they get conflated—when "I talked to two customers and they said" carries the same weight as a structured micro study with quality gates—the organization has an evidence standard problem.

Making this distinction explicit is not about hierarchy. It is about appropriate confidence and appropriate use. Anecdotal input is useful for generating questions. It is not useful for answering them. Expert judgment is useful for interpreting findings in context. It is not useful as a substitute for evidence. Research evidence is useful for informing decisions with bounded confidence. It is not useful as proof or certainty. Each type of input has a role, and the evidence standard defines what role each plays.

Another standard worth codifying is how research findings are cited and reused. Findings from a micro study have an expiry date, so when someone cites those findings in a planning document three months later, the questions are straightforward: Is the citation still valid? Has the expiry condition been met? Did the product change in ways that invalidate the original finding? Without a standard for citation hygiene, old findings accumulate in the organization's decision-making like sediment—referenced because they exist, not because they are still true. A simple rule—something like "findings older than their expiry condition must be verified before being cited"—prevents this. It is hard to enforce perfectly, but it is still worth stating.

A third standard concerns what happens when findings from different studies conflict. A micro study says users understood the pricing page; a sprint study three weeks later says users were confused by the pricing toggle. Both are credible. Both are bounded. They are studying different aspects of the same thing, but to a stakeholder reading both readouts, they look contradictory. The evidence standard should address this: newer findings supersede older ones for the specific aspect studied. Broader studies provide more context than narrower ones. Conflicting findings are a signal for further investigation, not a reason to cherry-pick the one that supports the preferred direction.

Rolling Out Governance Without Becoming the Research Police

Having the governance system designed is one thing. Getting the organization to adopt it is another—and it is where most governance efforts stall. The system described in this chapter—quality gates, conditional self-serve, escalation trees, evidence standards—looks clean on paper. In practice, rolling it out requires sequencing, sponsorship, and a tolerance for imperfection.

Do not launch everything at once. The fastest way to kill a governance system is to show up with a ten-page policy document and ask the organization to comply: nobody reads it, nobody follows it, and the researcher becomes the person who made everything harder—so the next request routes around you entirely.

Start with one team. Pick a product team that already requests research regularly and has a PM who is genuinely interested in getting better signal. Introduce the intake fields first: decision, decision owner, question, timeline. Do not call it governance. Call it scoping. Run a few studies through the process. Let the PM experience how the scoping conversation produces better results. Let them see how the boundaries section in the readout protects the findings from being misused. Let the quality gates be invisible at first. You are applying them, but the PM does not need to see the checklist yet—they just see that the readout is credible and defensible.

After two or three studies, introduce the gates explicitly. Show the PM the checklist. Explain that this is what makes the findings trustworthy, and ask whether they would be comfortable applying the same standard to studies they run themselves. Most PMs who have seen the difference between gated and ungated work will say yes. Now you have your first self-serve candidate and your first governance advocate—which matters more than any policy document.

Expand from there. The second team sees the first team's process and asks for the same thing. The researcher brings the same sequence: intake fields first, quality gates second, conditional self-serve third. Each team that adopts the system becomes a proof point for the next team. This is slower than a top-down mandate, but it is more durable because adoption is driven by demonstrated value rather than compliance pressure.

Leadership sponsorship matters, but the timing matters more. Do not ask for leadership buy-in before you have proof points. A research lead walking into a VP's office saying "I want to implement a governance framework for research quality" will get a polite nod and no action. A research lead walking in saying "three product teams are now running their own micro studies using this system, here are the quality standards they follow, here is an example of a study that caught a bad finding before it reached a decision, and I want to make this the standard across the organization" will get actual support. Show the results first. Ask for the mandate second.

One practical detail that accelerates adoption: make the templates easy. The intake form should take two minutes to complete. The prompt templates should be copy-and-customize, not build-from-scratch. The quality gate checklist should fit on one page. The readout template should have the four sections pre-labeled: findings, evidence, boundaries, expiry. Every piece of friction you remove from the process increases the likelihood that people will actually use it. Governance that requires effort gets worked around; governance that is easier than the alternative gets adopted.

When Governance Breaks Down

It will—not because people are malicious but because the incentives in product organizations favor speed over process. A PM will skip the researcher review and share findings directly because the meeting is in an hour. Someone will cite a four-month-old micro study in a planning document without checking the expiry. A designer will run a self-serve study with no evidence chains and present the AI summary as findings to a VP who does not know the difference. These are not hypothetical failures. They are predictable ones, and how you respond to them determines whether the governance system survives or becomes decoration.

The first principle is to treat breakdowns as system failures, not personal failures. If a PM skipped the review step, ask why. Was the review turnaround too slow? Was the PM unaware of the requirement? Was the study time-sensitive in a way the process did not accommodate? The answer tells you whether the system needs adjustment or the PM needs coaching. If three PMs skip the review in the same month, the system has a design problem. Fix the system. Do not send a reminder email about compliance.

The second principle is to correct publicly when possible, privately when necessary. If expired findings show up in a planning document, the correction should happen in the same forum—not as a callout but as a professional update. "These findings are from the March study of the previous pricing page. The pricing page has since been redesigned, so these findings no longer apply. If we want current data on the new page, I can scope a micro study." That is not policing. It is evidence hygiene, and it teaches everyone in the room that findings have shelf lives. If a PM shares ungated findings and they happen to be wrong, the correction is a private conversation first: "I noticed the readout did not include disconfirming evidence. When I checked the transcripts, there were two participants who had a very different experience. That changes the finding. Can we revise before it goes wider?" You are protecting the PM, not embarrassing them. Most people respond well to this when it is framed as collaboration rather than correction.

The third principle is to use failures as case studies. When a governance failure produces a visible consequence—an expired finding that led to a bad decision, a self-serve study that missed a critical usability problem because nobody checked the transcripts—document it and, with appropriate discretion, use it to make the case for the system. Not as a blame exercise but as a learning moment. "Here is what happened when we skipped the review step. Here is what the review would have caught. Here is the cost of not catching it." One concrete example does more to reinforce governance than ten policy documents.

The fourth principle is to build recovery paths, not just prevention. The governance system should include a mechanism for catching problems after the fact, not just before. A monthly audit of recent readouts—even a quick scan of five or six—tells you whether the quality gates are holding. Are findings linked to evidence? Are boundaries stated? Are expiry dates included? If compliance is high, the system is working. If compliance is drifting, you can catch it early and address it before it becomes the norm. The audit takes thirty minutes. The alternative is discovering six months later that half of the "research" in the organization's decision record does not meet basic evidence standards.

A Maturity Progression

Not every organization needs the full governance infrastructure described in this chapter. Where you start depends on where you are. The maturity progression below is not a prescriptive ladder. It is a diagnostic: figure out where your organization is, and focus on the next level rather than trying to build everything at once.

Level one is the researcher-only stage, where one researcher runs all studies, quality is a matter of personal habit, intake is informal, and demand management is reactive. This is where most organizations start when they adopt micro research. It works when the volume is low and the researcher has capacity, but it breaks when demand grows, which it will.

Level two introduces structured intake and service levels, with the researcher implementing a lightweight intake system in which requests are filtered before they consume research time, service levels are published, and prioritization is principled. The researcher is still the only one running studies, but the system around them is organized enough to handle moderate demand. Most organizations can get here within a few months of adopting micro research, and many can sustain this level indefinitely if the demand stays manageable.

Level three enables conditional self-serve, where non-researchers are running micro studies under defined conditions: approved prompt templates, quality gates as policy, and researcher review of readouts. The escalation tree determines what is self-serve eligible and what requires researcher ownership. The researcher's role shifts from running every study to running the complex ones and reviewing the simple ones. This level requires the governance infrastructure described in this chapter. Without it, level three collapses into ungoverned distribution. It also requires acknowledging that somebody needs to maintain the infrastructure: updating prompt templates, managing the intake pipeline, tracking compliance, coordinating reviewer availability, and keeping the escalation tree current. In smaller teams, the researcher absorbs this work. In larger organizations, this is research operations, and trying to run level three without dedicated operations support is how researchers burn out while the system they built slowly degrades around them. If the volume of self-serve studies exceeds what one researcher can review alongside their own work, the organization needs to invest in operations capacity or accept that the governance system will erode.

Level four represents embedded practice, in which research—including micro and sprint studies—is a regular part of how the organization makes decisions. Multiple people can run studies. Quality standards are organizational culture, not just policy. Evidence standards shape how findings are cited, reused, and superseded.

The researcher's role is primarily design, quality assurance, and the research that requires sustained interpretive judgment: sprint studies, deep research, and evidence governance. Research operations is a defined function, not a side responsibility. Someone owns the intake system, the templates, the compliance tracking, and the infrastructure that makes the whole system run. This level takes time to build and requires sustained investment in both the governance system and the people who maintain it. Most organizations do not reach it accidentally. They reach it because someone deliberately built the infrastructure and someone else deliberately maintained it.

The progression is not linear, and not every organization needs level four. The right level depends on the organization's size, the volume of product decisions that benefit from user evidence, and the number of people who want access to research tools. What matters is knowing where you are and what the next level requires, so you can build toward it deliberately rather than being forced into it by demand pressure.

The Federated Model and the Intelligence Function

There is a shift happening underneath the maturity progression that most research functions make without realizing it, and naming it changes how you lead toward it.

A service model research function sits downstream of decisions and responds to them; its value is proportional to the studies it runs and the findings it delivers. When demand exceeds capacity, it falls behind. When stakeholders stop asking, it becomes invisible. Most research functions were built this way—not because it is the right model but because it is the default. Nobody decides to build a service function. A service function is what you get when nobody has decided what else to build.

An intelligence function sits upstream of decisions and shapes the conditions under which they are made. It does not wait for questions. It maintains the organizational knowledge that makes better questions possible. The Frame is the artifact of this positioning. It does not exist because someone requested it. It exists because the research function built it, maintains it, and treats it as an organizational asset. The coverage matrix is not demand management. It is the intelligence function making gaps in organizational knowledge visible before anyone thinks to ask. The governance system is not bureaucratic overhead. It is the intelligence function setting the standards by which the entire organization produces and uses evidence.

The organizational model that makes this work at scale is the federated model, and it is the structure the Research Engine was designed to operate within.

The federated model has three layers.

The first is the central intelligence function, which is small—probably smaller than most research leaders expect—and has a mandate that looks different from the traditional research director role. This layer owns the Frame. It runs the deep research that builds and maintains the foundational understanding of users that the fast modes operate within, and sets the evidence standards that apply across the organization. It reviews the Frame-level question log and uses it to trigger deep research cycles before anyone requests them. It does not take requests and deliver findings; instead, it owns organizational intelligence as an asset and invests in it proactively.

What this requires from its leader is worth stating plainly. It is not managing a team of researchers who each own a product surface; it is owning the organizational layer that makes evidence usable across the whole system. It means treating the Frame as an asset that depreciates without active investment, the way a codebase degrades without maintenance, and it means setting standards that hold across a distributed research network without direct control over every researcher within it. This is closer to what a head of data or a chief analytics officer does than to what a traditional research director does, and most research leaders have not been prepared for it. That is worth knowing before the transition happens rather than after.

The second layer is embedded researchers operating as access points to the central intelligence layer rather than as independent research functions. An embedded researcher who is independent of a central function defaults to optimizing for their team. Their incentives align with the team's incentives. Over time, under consistent local pressure, their evidence bar drifts toward whatever the team will accept. An embedded researcher who operates as an access point has a different job. They are the interface between the team and the accumulated organizational knowledge. They bring Frame knowledge into planning conversations. They run micro and sprint research that feeds back into the Frame rather than staying siloed. They escalate questions that exceed the team's mandate rather than letting the team answer them with whatever is fastest.

The third layer is research operations functioning as governance infrastructure rather than logistical support. In a federated model, where micro research is distributed across embedded researchers and conditionally available to non-researchers, the operational layer that governs tool standards, methodology consistency, and knowledge management becomes the infrastructure that prevents the distributed system from fragmenting. Without it, each embedded researcher builds their own approach and their own quality bar. The knowledge they produce does not accumulate. The Frame cannot be maintained from inconsistent inputs. ResearchOps is what keeps the distributed system coherent enough to function as a single intelligence layer.

These three layers map onto the Research Engine as follows. The central function owns the Frame and deep research. The embedded layer runs micro and sprint research and contributes findings back to the Frame. ResearchOps governs the tooling and standards that make the embedded layer consistent enough to contribute usefully. The routing logic and the Decision Contract apply at every point where a team interacts with the research function, ensuring fast work stays within its mandate and questions that exceed it get escalated rather than quietly answered with whatever is available.

The federated model is not a prescription for every research function. A team of one does not need three layers. The maturity progression earlier in this chapter tells you what level of infrastructure you actually need given your size and volume. What the federated model describes is what a mature research function looks like when the scaling problem becomes real.

The transition from service to intelligence does not happen through a strategy document or a reorg. It happens through a consistent pattern of behavior: maintaining the Frame when nobody asked you to, making knowledge gaps visible before anyone flagged them, setting evidence standards that govern how the whole organization uses research, and holding the routing logic when the pressure is to call everything micro and ship it fast. The organizations that complete this transition are not the ones that argued for it in a planning cycle. They are the ones that built the system, demonstrated what it produced, and let the repositioning follow from the evidence.

The Research Engine is the operational model for an intelligence function. The three modes—the routing logic, Decision Contract, and governance system—are not tools for running a better service model. They are the architecture of a function that has decided its job is not to answer questions but to build and maintain the organizational intelligence that makes better questions possible.

Where This Goes Next

Demand and governance are two sides of the same organizational challenge. Demand creates pressure to produce more research faster, while governance ensures that what gets produced is credible. Together, they form the infrastructure that allows AI-assisted research to scale beyond one researcher running studies alone.

But there is a risk embedded in all of this scaling. Micro studies are fast. Sprint studies are responsive. Both modes fill calendars, generate visible output, and make the research function look productive. And both modes, if left unchecked, will cannibalize the slowest and most valuable kind of research: deep research.

Deep research does not fill calendars, does not generate two-day turnarounds, and does not produce readouts that can be dropped into planning meetings. Instead, it produces the foundational understanding that makes everything else work. And it is also the first thing to disappear when demand for fast work grows—not because anyone explicitly decides against it but because nobody makes the affirmative case for it while the micro requests keep coming.

Protecting deep work from the gravitational pull of fast work is the subject of the next chapter.

CHAPTER 10

Protecting Deep Research

Chapter 4 introduced the claim and Chapter 5 developed it fully: deep research builds the Frame, and everything else operates within it. Without a deep understanding of the domain, users, and problem space, the tight questions that micro research answers will be the wrong tight questions. You will be optimizing with precision inside a Frame that was never validated.

Most people nod when they read that, but most organizations ignore it in practice. This is not deliberate—nobody holds a meeting and decides to cancel deep research. What happens is subtler and more damaging. Micro research is fast. Sprint research is responsive. Both generate visible output, produce deliverables that can be dropped into planning meetings, and make the research function look productive. Deep research does none of those things. It is slow, generates ambiguous intermediate findings, does not produce two-day readouts, and lacks a clean ROI story. That is why it is the first thing to disappear when demand for fast work grows.

This chapter is about preventing that disappearance—not by arguing that deep research is important, which everyone already agrees with in principle, but by building the organizational and personal practices that protect it from the gravitational pull of fast work.

When Deep Research Is the Right Mode

The routing logic from Chapter 4 gives the abstract answer: high ambiguity, high stakes, long expiry. But deep research also has specific triggering conditions worth naming concretely, because the abstract answer does not always make the call obvious in practice.

The first is new territory. When the product moves into a new problem space, a new user segment, or a new market, the Frame does not extend automatically. Running sprint and micro research in new territory without first building foundational understanding produces precise answers to questions that may not be worth asking. Deep research comes before evaluation in new spaces—not as a phase with a formal end date, but as a recognition that the organization needs to understand the domain before it can ask the right questions about it.

The second is a broken or degraded Frame. Chapter 5 described the symptoms: the team keeps getting surprised by user behavior that should not happen given its model of the user; micro studies produce findings that contradict each other for no obvious reason; features launch with strong comprehension results and fail to drive adoption. When those symptoms accumulate, no amount of micro or sprint research fixes them. Those modes test within the Frame. Deep research steps outside it.

The third is a strategic decision carrying Frame risk. A product team about to make a significant investment based on an understanding of users that comes from research done three years ago is carrying risk, whether it recognizes it or not. Deep research before the investment is cheaper than the investment itself if the foundational assumptions turn out to be wrong.

The fourth is the Frame-level question log. Chapter 5 introduced this practice: a running record of the moments in fast research where something does not add up, where participants seem to be operating in a reality the team has not mapped. When that log accumulates three or four questions pointing at the same territory over a quarter, it signals a deep research project waiting to happen, with the log serving as the supporting evidence.

How Cannibalization Happens

Deep research does not get killed; it gets crowded out. Understanding the mechanism matters, because the mechanism is what you need to interrupt.

It starts with the demand explosion from Chapter 9. Micro research is adopted, stakeholders discover that they can get directional answers in two days, and requests increase. The researcher's calendar fills with micro studies—each one legitimate, each one fast, and each one connected to a real decision. The calendar looks full and productive, and the researcher feels busy and useful.

Then a sprint study comes in. It needs two weeks. The researcher looks at the calendar and sees micro studies stacked through the next three weeks. They could defer a few micro studies to make room, but each of those studies has a stakeholder waiting, a decision pending, and a timeline that will pass if the study does not happen now. The sprint study gets deferred. It eventually happens, but it is compressed into a shorter timeline than it needed, because by the time there was a gap in the calendar, the decision it was meant to inform has moved closer.

Deep research never even gets to the deferral stage. Deep research does not arrive as a request with a deadline. It arrives as a vague sense that the team does not understand something well enough—"We should really spend some time understanding how these users think about this problem." That sentence gets said in a meeting. Everyone agrees, but nobody schedules it. Three months later, the same sentence gets said again; everyone agrees again, and the micro studies keep running.

The cannibalization is not malicious. It is gravitational. Fast work has deadlines, stakeholders, and visible output. Slow work has none of those forcing functions. In the absence of deliberate protection, the work with forcing functions always wins—always. Deep research loses not because anyone decided it was less important but because it never had a deadline to protect it, and in organizations, work without deadlines does not happen.

The Cost of Running Without Deep Research

Deep research is the only mode that builds and rebuilds the Frame. When it stops, three things break—and they break in order.

The first to break is the Frame itself. Sprint and micro research continue running inside a Frame that is getting progressively less accurate. Each individual study is executed well, and the findings are technically correct. But the questions being asked are increasingly the wrong questions, because the Frame that determines which questions are worth asking is no longer current. The research looks productive, and the function looks healthy. The dysfunction only becomes visible downstream—when products ship and fail, when adoption is flat despite strong comprehension results, and when the team keeps getting surprised by user behavior that should not happen given its model of the user.

The second thing to break is coverage in territory the organization has never mapped. Frame degradation is at least visible in retrospect: contradictions accumulate, surprises mount, and something eventually fails. But the spaces the organization has never explored produce no signal. There are no contradictory findings because there are no findings. There are no surprises because nobody asked the questions. The product team makes decisions about user populations it has never studied, problem spaces it has never mapped, and contexts it has never observed—and the absence of research is invisible because nothing exists to flag it as a gap. The coverage matrix from Chapter 5 is the mechanism for surfacing this: areas marked as low coverage or operating on assumption are the spaces where deep research has never happened. Without deep research, those spaces do not get smaller. They get larger as the product expands, as new segments emerge, and as the market shifts into territory the team has no foundational understanding of. The organization does not know what it does not know, and without deep research, it never finds out until a product decision walks into the dark and fails there.

The third thing to break is the signal loop. Fast research generates Frame-level questions constantly: participants say things that do not fit the team's model, and findings point at something bigger than the question being asked. The Frame-level question log fills up. Without deep research, nothing answers those questions, and the log becomes a record of mounting uncertainty that nobody has the mandate or the time to address. When the Frame is current, contradictory fast research findings are signals worth investigating. When the Frame is degraded, those same contradictions become background noise, because so much of what the team finds seems slightly off that no single contradiction stands out as worth escalating. The team normalizes the confusion rather than treating it as a system problem.

That normalization is the most expensive outcome—not because it produces bad research but because it produces a research function that looks fully operational while the foundational understanding it depends on quietly expires.

Fast Research as a Deep Research Trigger

One of the most useful things micro and sprint research can do—and one of the least recognized—is signal when deep research is needed.

The signal pattern is consistent. You run a micro study on a pricing page: the comprehension results are fine, but participants say things that suggest they think about pricing differently than the team assumes. You note it, but the micro study was about comprehension, not mental models, so it goes into the parking lot. You run another micro study on the onboarding flow: the flow tests fine, but participants describe their goals in terms the team does not use and does not understand—another parking lot note. A sprint study on a new feature concept surfaces strong reactions, but those reactions are grounded in context the team did not anticipate. The participants are not wrong. They are operating in a reality the team has not mapped.

Individually, each of these is a finding within a study. Collectively, they are a signal that the Frame needs updating. The team's understanding of the users—their context, their mental models, their priorities—is outdated or incomplete. The fast research is doing its job within its boundaries. But the boundaries keep bumping into something bigger.

The discipline here is to track these signals rather than letting them evaporate. A simple mechanism is to maintain a running log of Frame-level questions that surface during micro and sprint studies—not the findings themselves but the questions those studies could not answer because they were outside scope. For example: "Participants seem to think about pricing as a monthly budget decision, not a feature comparison. We assumed feature comparison. Do we understand how this segment actually thinks about software spending?" That is a deep research question generated by a micro study. If the log accumulates three or four questions in the same territory over a quarter, that is a deep research project waiting to happen.

This reframes the relationship between fast research and deep research. They are not competing for the same calendar space. They are part of the same system. Fast research produces answers and generates questions. Deep research answers the questions that fast research generates. The fast work feeds the slow work, and the slow work corrects the Frame that the fast work operates within. Without both, the system degrades.

They are part of the same system. Fast research produces answers and generates questions. Deep research answers the questions that fast research cannot. This is the research engine from Chapter 4 in action, with the Frame from Chapter 5 operating as designed. Deep research builds the Frame. Micro and sprint operate within it. Micro and sprint signal when the Frame is breaking. Deep research rebuilds. The organizations that protect deep research are not being precious about methodology. They are maintaining the engine that makes everything else work.

How Deep Research Connects Back to the System

Protecting deep research is only half the argument. The other half is making sure that when deep research does happen, it connects back into the system rather than producing a deliverable that influences one road map cycle and then sits in a repository.

The connection happens in three ways.

The first is the Frame update. When deep research is complete, the structured document gets updated, the coverage matrix changes, and the Frame steward owns the transition from old belief to new belief. This requires explicitly retiring what was previously believed in areas the deep research has covered, marking it as superseded, and stating what replaced it and why. The Frame-level question log is then reviewed: which questions did the deep research answer, which remain open, and which generated new questions for the next cycle.

The second is the recalibration of fast research. Sprint and micro studies that were designed against the old Frame may need to be reframed. Questions that seemed important before the deep research may no longer be the right questions. New questions that the deep research surfaced become candidates for the sprint queue. Deep research does not just answer questions. It reorders the question queue.

The third is the signal loop forward. Deep research generates its own Frame-level questions as a byproduct—things the researcher encountered that the study was not designed to answer but that still matter. These go into the Frame-level question log, not as findings but as signals that inform the next planning cycle. The cycle continues: fast research signals what the Frame does not cover; deep research rebuilds or extends the Frame; the new Frame enables better fast research, which generates new signals that eventually trigger the next deep research cycle.

When that cycle is running, the research function is not just producing studies. It is producing organizational understanding that compounds over time. The fast modes become more useful as the Frame becomes more accurate, and the deep research becomes more focused as the fast modes get better at signaling where the Frame is breaking. The system begins to self-correct.

Making the Case for Slow Research in a Fast Culture

Protecting deep research requires more than a good argument. It requires an argument that works in the specific organizational context you are operating in. Saying "deep research is important" is not enough. You need to connect deep research to something the organization already cares about.

The most effective case I have found ties deep research to risk reduction. Organizations understand risk. They spend significant effort and money managing risk in engineering, compliance, finance, and operations. Research risk—the risk of building the wrong thing because the team misunderstands the users—is the same kind of risk. It just does not have an obvious price tag.

The pitch goes like this. Micro and sprint research reduce execution risk. They help the team make better decisions about specific screens, flows, features, and pricing. They are fast, and they are valuable. But they do not reduce foundational risk. Foundational risk is the risk that the team's understanding of the users, the market, or the problem space is wrong. That risk accumulates silently: the team ships features that test well in comprehension studies but fail to drive adoption; the team enters a market with assumptions that turn out to be incorrect; the team builds a product around a mental model that users do not share. These are not execution failures. They are Frame failures—and the only way to reduce Frame risk is deep research.

When you Frame it as risk, the conversation changes. You are not asking for permission to do slow, expensive, hard-to-justify work. You are identifying a category of organizational risk that is currently unmanaged and proposing a mitigation. That is a conversation any business leader can engage with.

Another angle that works is to connect deep research to decisions the organization knows it needs to make. If the product team is planning to enter a new vertical next year, deep research on that vertical is not speculative. It is preparation for a specific strategic move. If the company is redesigning its core pricing model, deep research on how users think about pricing is not academic. It is risk reduction for a high-stakes decision. The more tightly you can connect deep research to named strategic priorities, the easier it is to protect.

What does not work is arguing for deep research on the basis of research best practices. "Good research functions do deep research" is a craft argument: it persuades researchers, but it does not persuade PMs, VPs, or CFOs. They need a business argument, and the business argument is risk.

Sometimes the case lands and nothing happens. The VP agrees that deep research is important, the planning document includes a line about foundational research, and then the quarter starts, the micro requests arrive, and nobody blocks the time. The agreement was real, but the follow-through was not. This is not bad faith—it is how organizations work. Agreement in planning does not automatically produce allocation in execution.

When this happens, the researcher has a few options, and none of them are comfortable. The first is to run deep research small. If the organization will not fund a six-week project, run a three-week project. If three weeks is too much, run a one-week intensive—four or five extended sessions with users in the target segment, enough to start seeing the Frame even if you cannot map it completely. Compressed deep research is not ideal, but it is better than no deep research. It produces enough signal to demonstrate value, which makes the case for the next effort easier.

The second option is to embed deep research in sprint research. A sprint study can be designed to include a few sessions that go deeper than the sprint question requires. You are officially running a sprint study on pricing evaluation, but two of your follow-up sessions spend extra time on how the participant thinks about software spending in general, which is a deep research question. The sprint deliverable covers the sprint question. The deep research signal goes into your Frame-level question log. This is not a substitute for dedicated deep research. It is a way to gather deep research signal when dedicated time is not available.

The third option is to let the consequences become visible. If the organization will not invest in deep research, the Frame will eventually break in a way that is visible to leadership: a product launch that fails despite strong micro-study results, a market entry that misreads target users, or a strategic bet that does not pay off because the foundational assumptions were wrong. When that happens, the researcher who has been tracking Frame-level questions and flagging the need for deep research has the credibility to say "this is what I was trying to prevent." That is not satisfying, and it is not fast. But it is sometimes the only path to organizational learning on this point. Some organizations need to feel the cost of a missing Frame before they will invest in building one.

Calendar Protection

The case for deep research gets you the permission. Calendar protection gets you the time.

The fundamental problem is that deep research requires extended blocks of time that are incompatible with the reactive cadence of micro research. A micro study occupies two to three days. Deep research requires sustained engagement over weeks. If the calendar is managed reactively, micro studies fill every gap, and deep research never gets a contiguous block. You need to protect the time before it gets consumed.

The simplest protection mechanism is to schedule deep research like you would schedule any other commitment. If you have a deep research project planned for a quarter, block the weeks on your calendar now—not with vague promises such as " "I will find time later," but with actual, concrete time blocks. If someone asks for a micro study during those weeks, answer with: "I am running a deep research project during that period. I can take your micro study the week after, or we can discuss whether someone else can run it." The deep research project has a calendar commitment just like a sprint study; it is real work with real time allocated, not a residual that happens if there are leftover hours.

This sounds simple. In practice, it is the hardest discipline in this entire book. The pressure to give up deep research time for micro studies is constant and comes from every direction: a stakeholder has an urgent question; a VP needs something right away; a product launch has moved up and the team needs three micro studies in two weeks. Each individual request is reasonable. Each one, if you accommodate it by cannibalizing deep research time, reinforces the pattern that deep research is the flexible resource that gets sacrificed whenever something urgent appears.

The mitigation is to treat deep research time as a named commitment with a stakeholder, not as discretionary time. Deep research should have a sponsor—a product leader, a VP, someone with organizational authority who has agreed that this work matters. When the pressure comes to sacrifice deep research time for a micro study, the conversation is not "can I skip my deep research work"; it is "this micro study would require me to delay the deep research project that [sponsor name] and I agreed to. Let me check with them before making that trade-off." The sponsor provides organizational cover. Without it, the researcher is making a personal choice to protect slow work over fast work, and that choice is hard to sustain under pressure. With it, the researcher is honoring a commitment that someone with authority endorsed.

Another protection mechanism: pair deep research with deliverable milestones. Deep research is vulnerable because it does not produce visible output on a predictable cadence: a micro study produces a readout every few days, while deep research produces understanding over time. The lack of visible progress makes it easy to deprioritize. Counter this by building intermediate deliverables into the project—a stakeholder update at the end of week two, a preliminary findings document at the halfway point, a synthesis workshop with the product team before the final deliverable. These milestones make the work visible, which makes it harder to cancel.

The Deep Research Budget

Calendar protection is the tactical mechanism. The deep research budget is the strategic one.

A deep research budget is not a financial budget, although it can be. It is a time commitment: a percentage of the research function's capacity that is reserved for deep research work regardless of micro and sprint demand. The percentage is not fixed. It depends on the organization's maturity, the stability of the product, and the pace of strategic change.

An organization with a mature product in a stable market might allocate 10 to 15 percent of research capacity to deep research. The Frame is well established, and micro and sprint research can operate within it productively. In this context, deep research is maintenance: checking that the Frame still holds, updating it when the market shifts, and exploring adjacent spaces when the product road map looks outward.

An organization entering new markets, undergoing strategic pivots, or building new product categories needs significantly more—30 percent or more. The Frame is not established, and running micro and sprint research without a Frame produces precise answers to potentially wrong questions. Deep research is not a luxury in this context. It is a prerequisite for everything else.

An organization in between—one that has a stable core product but is expanding into new areas—might split the difference. Protect enough deep research capacity for the new areas while maintaining micro and sprint coverage for the core.

The specific numbers matter less than the principle: deep research capacity is a named allocation, not a residual. It is decided in advance, by the research lead and their leadership, as part of planning. It is not what is left over after micro studies take their share. If it is treated as a residual, it will be zero. That is the gravitational pull in action.

When Deep Research Gets Squeezed Anyway

Even with calendar protection, a named sponsor, and a deep research budget, there will be moments when deep research gets squeezed—a critical product launch, a crisis, or a quarter where the demand for fast work is genuinely overwhelming and something has to give. The question is not whether this will happen. It is how you respond when it does.

The first response is to acknowledge it explicitly rather than letting it happen silently. "I am pausing the deep research project for two weeks to handle the launch-related micro studies. The deep research project will resume on [date], and this delays the final deliverable by two weeks." That communication does two things: it makes the trade-off visible so that the organization knows deep research was sacrificed and for what, and it creates a commitment to resume. That commitment is much harder to ignore than a vague intention to "get back to it when things calm down," because things do not calm down—they get replaced by the next thing. A specific resume date is the only reliable mechanism.

A note on what deep research produces, because the absence of a defined deliverable is part of what makes deep research vulnerable. Micro research produces a directional readout. Sprint research produces a short report or focused deck. Deep research produces understanding, which is real but hard to put on a calendar as a milestone.

In practice, the most effective deep research deliverables I have seen take one of three forms. The first is a reframed problem statement: a document that articulates how the team's understanding of the users or the problem space has changed, what the previous assumptions were, and what the updated Frame looks like. This is typically two to five pages and becomes a reference document that micro and sprint studies are designed against. The second is a mental model or journey framework: a visual or narrative artifact that captures how users think about the domain, what their actual workflow looks like, and where the product fits or does not fit in their reality. This is the kind of deliverable that gets pinned to a wall or referenced in planning for months. The third is a strategic recommendation tied to specific evidence: not "users want X" but "users think about this domain in this way, which means our current approach assumes Y, and the evidence suggests Y is wrong. Here are the implications for the road map."

All three share a characteristic that distinguishes them from micro and sprint deliverables: they do not expire quickly. A micro readout expires when the screen changes. A deep research deliverable remains valid until the market shifts, the user base changes, or the problem space evolves. That longer shelf life is part of the value, and it is part of the case for the investment.

The second response is to track the squeezes. If deep research gets paused once a quarter for a product launch, that is a reasonable trade-off. If deep research gets paused every three weeks for urgent micro studies, the deep research budget is not real. It is aspirational. The squeezes are the data that tells you whether the protection mechanisms are working or failing. If they are failing, the capacity conversation from Chapter 9 applies: the organization is underinvesting in research capacity relative to the demand it is generating, and the shortfall is being taken from the work that matters most.

The third response is to resist the temptation to compress deep research rather than pause it. When deep research time gets squeezed, the instinct is to make the deep research project faster: fewer sessions, shorter engagement, less analysis. This produces deep research that looks complete but is not. Compressed deep research is just an expensive sprint study with a different label. If the deep research project cannot be done well in the compressed timeline, it is better to pause and resume than to produce shallow findings and call them foundational. Deep research that is not deep enough to update the Frame has not done its job.

The Relationship Between the Three Modes

Throughout this book, micro research, sprint research, and deep research have been presented as distinct operating modes with different scopes, timelines, and purposes. But in practice, they are not separate activities. They are parts of a single system.

Deep research builds the Frame: the foundational understanding of users, their contexts, their mental models, and their needs. Sprint research explores specific questions within the Frame, going deep enough to understand how users experience particular flows, features, or decisions. Micro research tests specific artifacts within the Frame, producing fast directional signal on narrow questions. Each mode depends on the ones above it. Micro research without sprint is narrow and disconnected. Sprint research without deep research is exploring within a Frame that may be wrong. Deep research without micro and sprint is foundational but disconnected from the daily decisions that shape the product.

The healthiest research functions I have seen treat the three modes as a cycle rather than a hierarchy. Deep research produces the Frame. Micro and sprint research operate within it and, over time, surface signals that the Frame needs updating. Those signals trigger new deep research. The Frame gets updated, enabling better micro and sprint questions, and the cycle continues.

When the cycle breaks, it breaks at deep research. Micro and sprint keep running because they have deadlines, stakeholders, and visible output. Deep research stalls because it has none of those things. The Frame gets stale, and micro and sprint questions start hitting walls. Stakeholders notice that research findings are less useful than they used to be but cannot articulate why. The reason is that the Frame expired and nobody updated it.

Protecting deep work is ultimately about protecting the cycle. It is not about valuing one mode over the others. It is about recognizing that the fast modes depend on the slow mode, and that the slow mode will not happen unless someone makes it happen. That someone is the researcher. It is part of the job—not the most visible part, and not the part that generates two-day readouts and stakeholder gratitude, but the part that makes everything else work.

Where This Goes Next

This chapter argued that deep research needs active protection because the gravitational pull of fast work will cannibalize it otherwise. Calendar protection, named sponsors, deep research budgets, explicit pauses and resumes, and the use of fast research as a deep research trigger are all mechanisms for ensuring that the slowest mode survives in an environment optimized for speed.

The final chapter steps back from operations entirely. It looks at where the research craft is heading: what gets easier as tooling improves, what remains difficult regardless of tools, and how the researcher's role evolves when execution is no longer the bottleneck. It is a chapter about the future, written for practitioners who will be living in it.

CHAPTER 11

The Future of the Craft

This book has been about an operating model: how to route research requests, run micro and sprint studies with AI-assisted tools, maintain quality at speed, manage demand, govern at scale, and protect the deep work that makes everything else possible. If you have read this far, you have a system. The question this final chapter addresses is what happens to the system, and to the researcher, as the tools continue to improve.

I am not going to predict specific tools because the tools that exist now will be replaced or transformed within a few years. Outset, which I have referenced throughout this book, may look completely different by the time you read this, and new tools will emerge that do things none of us have anticipated. Making specific predictions about tooling is a losing game, and I am not interested in playing it.

What I am interested in is the structural question. As AI tools handle more of the execution work that researchers used to do, what becomes the researcher's actual job? What skills become more valuable? What skills become less relevant? Where does the craft go when the bottleneck shifts permanently from execution to judgment?

What Gets Easier

The trajectory is clear even if the specific tools are not: execution is getting easier and will continue to do so.

Recruitment is getting faster. Tools that connect researchers to participants, screen for eligibility, and schedule sessions are compressing into hours a process that used to take days. Panel quality remains a problem, as Chapter 8 discussed, but the logistics of finding and scheduling participants are steadily becoming less of a bottleneck.

Moderation is getting more capable. AI-moderated sessions today can follow scripts, ask follow-up probes, handle branching logic, and capture responses in multiple modalities. They are imperfect—they sometimes lead, they miss nuance, and they do not know when something unexpected is the most important part of the conversation—but they are better than they were a year ago, and they will be better still a year from now. For structured research tasks, the gap between what a skilled human moderator can do and what an AI moderator can do is slowly closing.

Transcription is effectively solved. Real-time, high-accuracy transcription with speaker identification and timestamps is available now; what was a meaningful time sink five years ago no longer is.

Initial summarization is also improving. AI-generated summaries of research sessions are useful first drafts today, though they miss nuance, smooth away contradiction, and optimize for clean narratives, all of which require researcher correction. Nevertheless, the quality of these summaries improves with each model generation, and the gap between the AI draft and the final synthesis is narrowing.

Pattern detection across large datasets is improving as well. When you have twenty transcripts from a micro or sprint study, AI tools can identify convergence and divergence across participants faster than a researcher reading each one sequentially. The patterns still need to be validated, but the initial detection—the "here are the themes that appear most frequently and here is where participants disagree"—is becoming faster and more reliable.

All of this means that the execution layer of research—the part that involves logistics, scheduling, moderating, transcribing, and generating initial summaries—is compressing. It took weeks, it takes days now, and it will take hours eventually. The time a researcher spends on execution is shrinking, and it will continue to do so.

What Remains Difficult

Everything that involves judgment remains difficult. The tools are not coming for judgment. They are coming for execution. And the distinction between the two is the most important thing a researcher can understand about the future of their profession.

Ambiguity remains difficult. Deciding whether a question requires micro, sprint, or deep research; recognizing when a seemingly simple question hides a complex problem; knowing when the data is telling you something unexpected versus when the data is noisy: these are judgment calls that require domain knowledge, experience, and the kind of pattern recognition that comes from years of doing the work. AI tools can present data. They cannot tell you what the data means in context.

Organizational politics remains difficult. Saying no to a VP; navigating a validation request that is really a political maneuver; building trust with stakeholders who have been burned by research that was slow, irrelevant, or both; and making the case for deep research in a culture that values speed are human skills in the deepest sense. They require reading people, reading rooms, and making strategic choices about when to push and when to accommodate. No tool handles this.

Interpretation risk remains difficult. A micro study produces a finding, but what does it mean for the product decision? How much weight should it carry, and what limitations does the stakeholder need to understand? How do you present mixed evidence without either overstating the consensus or drowning the decision maker in caveats? These are judgment calls that sit at the intersection of research methodology, product context, and communication skill. They are the highest-value work a researcher does, and they are the work that is least affected by tooling improvements.

Incentive misalignment remains difficult. Stakeholders who want research to confirm their existing beliefs; teams that celebrate positive findings and ignore negative ones; and organizations that treat research as a checkbox rather than an input to decision-making are structural problems that exist regardless of how good the tools are. A faster tool in a misaligned organization just produces confirmation bias more quickly.

Ethics and sensitivity remain difficult. Research that touches trust, safety, fairness, power dynamics, or vulnerable populations requires human judgment about what to ask, how to ask it, and how to handle what you hear. These are not execution problems. They are moral and professional judgment problems. AI tools should not handle them, and responsible researchers will not let them.

The pattern is clear: everything that requires a human to weigh context, navigate ambiguity, exercise judgment, or make ethical choices remains difficult. Everything that requires a system to execute consistently, quickly, and at scale gets easier. The researcher's value was always in the judgment. The tools are just making that fact impossible to ignore.

The Role Shift

When execution compresses, the researcher's role shifts—a change that has already begun, and which this book has described implicitly throughout. Now I want to name it explicitly.

The traditional researcher role was defined largely by execution: recruiting participants, writing discussion guides, moderating sessions, transcribing, synthesizing, and presenting. The researcher was valued for their ability to do these things well, and the skills that mattered were moderation technique, rapport building, note-taking, and the ability to synthesize qualitative data into coherent themes. These are real skills that are not going away, but they are becoming a smaller percentage of the job.

The emerging researcher role is defined by design and governance: designing the right question, designing the study that answers it, and designing the quality gates that ensure the answer is credible. It involves governing how evidence is produced, used, cited, and retired across the organization. The researcher becomes the person who shapes what gets studied, how evidence is evaluated, and what standards apply, rather than the person who runs every session personally.

This is the shift from practitioner to architect. The practitioner does the work, and the architect designs the system that produces the work and ensures its quality. Both are essential, but as tools take over more of the practitioner's execution tasks, the architect's skills become the differentiator.

Concretely, this means the skills that matter most for the next generation of researchers are question design (the ability to take a vague organizational need and turn it into a researchable question matched to the right mode and the right rigor level); evidence governance (the ability to set and enforce standards for how evidence is produced, evaluated, and used in decision-making); organizational influence (the ability to navigate stakeholder relationships, make the case for research investment, and protect the research function's integrity under pressure); and interpretive judgment (the ability to look at data, understand what it means in context, identify what it cannot tell you, and communicate all of that clearly).

Notice what is not on that list: moderation technique, transcription speed, or the ability to run five sessions in a day without losing focus. These skills were essential when the researcher was the execution engine, but they are less critical when tools handle execution. They remain useful, especially for follow-up sessions in sprint research and for deep research work, but they are no longer the core of the job.

To make this concrete, consider how the shift looks like in practice. A researcher in the traditional model might spend a typical week this way: scoping a study and writing a discussion guide on Monday, moderating six sessions on Tuesday and Wednesday, transcribing and starting synthesis on Thursday, and finishing synthesis and drafting a readout on Friday. The researcher is the execution engine. Every step requires their direct involvement. They touch the work from start to finish.

A researcher in the new model might spend the same week this way: doing question hygiene on two incoming micro requests and reviewing a self-serve readout from a PM on Monday morning; designing prompts for a sprint study and launching the AI-moderated phase on Monday afternoon; reviewing the first three transcripts from the sprint study, adjusting one probe, and delivering a micro readout from a study that completed overnight on Tuesday; conducting three follow-up sessions for the sprint study while a second micro study collects data in the background on Wednesday; synthesizing the sprint data, auditing AI summaries against transcripts, and delivering the second micro readout on Thursday; and spending Friday morning writing sprint findings and Friday afternoon on deep research work, reviewing the frame-level question log, and preparing for next week's extended user sessions.

It is the same week, and it results in more decisions informed by evidence with more studies in flight. However, the researcher's time is spent differently: there is less moderating, less transcribing, and less scheduling, and more designing, more auditing, more interpreting, and more governing. The execution compressed, and the judgment expanded to fill the space. That is the shift, not as theory but as a calendar.

The Output Shift

The role shift has a downstream consequence that most researchers have not yet confronted: the output has to change form.

The standard research deliverable is a finding: "Users are confused by the onboarding," "The pricing page does not communicate tier differences," or "The checkout flow breaks at step 3." These tell a team what is wrong, but they do not tell the system what right looks like.

Consider the difference between these two sentences: "Users feel overwhelmed by too many offers" is a finding; "The system should present no more than three offers per session, ordered by predicted relevance, with a suppression rule that withholds all offers for twenty-four hours after a failed redemption" is a behavioral specification. The same user knowledge underlies both, but they are completely different outputs. The first describes what users experienced, while the second prescribes what the system should do. The first requires a human to translate it before it can influence the product, whereas the second is ready to build from.

The readout was designed for humans who would interpret findings, carry them forward, and translate them into decisions using judgment and memory—imperfectly, but with flexibility. When a PM was the translation layer, stopping at the finding was lossy but functional. The PM absorbed the research, applied their context, and wrote the spec using their own judgment. The research reached the product through a human relationship. When that relationship was good, the bridge held.

As engineering organizations restructure around AI agents and spec-driven development, that translation layer is thinning. An agent reads a spec and builds to it; if the spec contains the behavioral constraint that research would have informed, the constraint gets built. If it does not, it does not exist. There is no meeting where an engineer pauses and asks whether research said something about this. The specification is the specification. User knowledge is encoded in it or it is not.

This does not mean every research output needs to be a behavioral specification. Foundational generative research, early-stage exploration, and strategic investigations produce understanding that is not yet specific enough to specify. The Frame is the repository of that understanding, and it feeds the specification when the time comes. Still, the direction of travel is clear: the researcher who treats the finding as the terminal artifact and the translation as someone else's job is betting on a translation layer that is getting smaller.

The output shift is uncomfortable for specific reasons. A finding can be nuanced, but a specification cannot. A finding describes ambiguity accurately, but a specification collapses ambiguity into a decision, and that decision can be wrong in a way that everyone can see. The entire output convention of UXR is organized around the finding precisely because the finding is hard to evaluate against outcomes. The specification is not. That is a different kind of professional exposure, and most researchers have not been asked to accept it.

There is something real that gets lost in this shift. The specification does not carry narrative. The story of the user who tried to redeem three offers, watched each fail, and said "I just felt stupid, like I was doing something wrong" is what makes a product team care. The spec is what makes the product change. Both matter, and the narrative survives as the reason behind the specification—the evidence that justifies the constraint. But it is no longer the primary output. The primary output is the behavioral constraint itself, specific enough to be tested, grounded in evidence, and ready for whatever system will implement it.

The researcher who produces that is not doing less than the researcher who produces the readout. They are doing more. The commitment is harder, the exposure is greater, and the contribution is direct rather than mediated. It requires exactly the skills that the role shift is moving toward: interpretive judgment, question design, and the ability to translate what users need into what systems should do.

The Identity Question

The role shift creates an identity question that many researchers are struggling with, and it is worth addressing directly rather than pretending it does not exist.

If you became a researcher because you love talking to people, because the interview is where the magic happens, because the moment a participant says something you did not expect is the best part of your job, the shift toward design and governance can feel like a loss. The thing you love doing is being automated. The thing you are being asked to do instead—shaping questions, reviewing readouts, setting quality standards, managing stakeholders—feels administrative compared to the human connection of a live interview.

I understand this. I also think it is worth interrogating. The magic of the unexpected participant response does not disappear in the new model. It still happens in sprint research follow-up sessions. It still happens in deep research. It still happens when you review a transcript and a participant says something that reframes how you think about the problem. What changes is the ratio: less time in sessions and more time in systems; less time listening and more time designing what to listen for.

Whether that trade-off is acceptable is a personal question. Some researchers will embrace the shift and find that their highest-value work, the work they are best at and most energized by, is the design and governance layer. They will become research leads, research directors, and evidence architects. Some researchers will resist the shift and seek roles that preserve the practitioner model: deep research–focused positions, research agencies, academic research, or organizations early enough in their research maturity that they still need a hands-on practitioner. Both paths are legitimate.

What is not a viable path is pretending the shift is not happening. The tools will continue to improve, and execution will continue to compress. The organizations that adopt AI-assisted research will need fewer practitioners and more architects. Researchers who build design and governance skills alongside their practitioner skills will have the most options, while researchers who define themselves entirely by execution skills will find the market for those skills contracting.

That is not a threat. It is an honest assessment. And I would rather say it plainly than let you be surprised by it.

What This Means for Hiring and Growth

If the role is shifting, the way organizations hire researchers and the way researchers develop their careers must shift with it.

For research leaders hiring in this environment, the traditional interview loop that emphasizes moderation technique and synthesis craft is testing for skills that are becoming less differentiating. A candidate who can run a beautiful interview but cannot scope a study, design quality gates, or navigate a stakeholder conversation about risk is a practitioner without the architect layer. That was fine when the job was primarily execution, but it is insufficient when the job is primarily design and governance. The interview should test whether the candidate can take a vague business question and turn it into a scoped study matched to the right mode, identify what an AI-generated summary missed, explain to a skeptical PM why a question needs sprint depth rather than micro speed, or write a readout with findings, evidence, boundaries, and an expiry date that holds up under scrutiny. These are the skills that determine impact in the new model.

This does not mean moderation skills are irrelevant. Sprint research follow-ups and deep research work still require a researcher who can build rapport, listen carefully, and follow unexpected threads. But moderation is one skill among several, not the defining one. Hire for the full range.

For individual researchers thinking about their own development, the highest-return investments are in the skills the tools cannot replicate. Get better at question design: practice taking vague requests and turning them into scoped studies until you can do it in a five-minute conversation. Get better at evidence governance: learn to set quality standards, review others' work, and build systems that scale beyond your own studies. Get better at organizational influence: learn to make the case for research investment, navigate political dynamics, and say no in ways that build trust rather than erode it. Get better at interpretive judgment: practice looking at data and articulating not just what it says but also what it means, what it does not mean, and what should happen next.

These skills are harder to develop than moderation technique because they are less procedural. There is no discussion guide template for organizational influence. There is no script for interpretive judgment. They develop through practice, mentorship, and the willingness to operate in ambiguity rather than retreating to the comfort of a well-structured study plan. They are also the skills that will define who leads research functions in the next five years and who gets managed out as organizations restructure around the new model.

Boundary Cases

Throughout this book, I have described a system for matching research modes to questions based on risk, ambiguity, and expiry. The system works for the vast majority of requests a research function encounters. But there are boundary cases—situations where someone asks you to do something fast that should never be fast—and the system needs to hold.

Here are the ones I think about most.

Someone asks for a micro study on a question that involves user safety. Safety questions are not micro, regardless of how simple they seem. "Do users understand this warning label" sounds like a comprehension test. But if the warning label is about a safety risk, and users misunderstand it, the consequences are not "the team revises the copy." The consequences are that users are exposed to a risk they did not understand. Safety questions need more rigor, more careful sampling, more thorough analysis, and more conservative interpretation than micro research provides. Route them to sprint at a minimum. Consider deep research if the safety context is not well understood.

Someone asks for a micro study on a question that involves a vulnerable population, such as users with disabilities, users in financial distress or health crises, or users who are minors. These populations require research approaches that prioritize participant well-being alongside data quality. AI-moderated sessions may not provide the sensitivity, the ability to pause when someone is distressed, or the ethical judgment to know when a question should not be asked. These studies need researcher involvement throughout. They are not self-serve candidates, and they are not micro candidates unless the question is truly narrow and the population-specific risks have been carefully considered.

Someone asks for a micro study to resolve a strategic disagreement between senior leaders. This is the political request from Chapter 7, but at higher stakes. Two VPs disagree about direction. One of them asks for research to settle it. The micro study runs, and the findings support one position. The losing VP questions the methodology, the sample, or the interpretation. The research function is now caught in a political crossfire it did not create and cannot resolve. Strategic disagreements need strategic research. Sprint at a minimum, with careful scoping that both parties agree to in advance. A micro study dropped into a political dispute is a weapon, not a tool.

Someone asks you to compress a deep research project into a sprint timeline because "we just need the highlights." Deep research does not have highlights. It has an evolving understanding that builds across extended engagement. Compressing it produces surface-level findings dressed up as foundational insight. The team members think they understand the users, but they understand only the first layer. The second and third layers, where the real insight lives, never got explored. If the timeline allows a sprint study, run a sprint study and call it that. Do not call it deep research. The label matters because it sets expectations about what the findings can support.

Someone asks for a study, any study, on a question where the answer cannot change the outcome. The product will ship on the scheduled date regardless. The road map is locked. The strategy is set. Nobody with authority will alter course based on what the research shows. This is validation theater, and the right answer is to decline, reframe, or at a minimum state explicitly in the readout that the findings were produced under conditions where the decision was already made. The research record should reflect reality, not the fiction that the study influenced something.

These boundary cases are where the system gets tested. The taxonomy, routing logic, MVR framework, decision contract, and quality gates: all the infrastructure described in this book exists so that when someone pushes you to do something fast that should not be fast, you have a principled basis for holding the line. The system absorbs the pressure so you do not have to.

In Closing

This book started with a structural observation: product teams generate decisions faster than research can inform them. The traditional research operating model does not match the tempo. The gap between decision velocity and evidence velocity is where bad decisions happen.

AI-assisted research is a real and meaningful response to that gap. It compresses execution. It enables micro and sprint studies that were not possible before. It increases the percentage of product decisions that are informed by user evidence rather than opinion. Those are genuine gains.

But the tools do not solve the problem alone. They need a system around them. A taxonomy for routing questions to the right mode. Quality gates for ensuring that fast research is credible research. Governance for scaling beyond one researcher. Protection for the deep work that makes everything else possible. And a researcher who understands that their value is not in running sessions but in designing what to study, judging what the data means, and governing how evidence is used.

The tools will keep improving. The execution will keep compressing. The specific platforms and features I referenced in this book will evolve or be replaced. That is fine. The system described here is tool-neutral. It is built on principles—routing by risk, matching rigor to stakes, maintaining evidence standards, protecting foundational research—that apply regardless of which tool is executing the sessions.

What does not change is the need for judgment: the need for someone who can look at a question and know what kind of treatment it deserves; who can look at data and know what it means and what it does not mean; who can look at an organization and know where the evidence gaps are and what to do about them; who can hold the line when the pressure is to move fast and the question demands that you move carefully.

That is the researcher's job. It always was. The tools just made it clearer.

"We stopped asking because we assumed you could not help fast enough." That sentence motivated this book more than any other. Not because it was an indictment of research but because it was an indictment of an operating model that made research invisible at the moments it was needed most.

The system in this book is designed so that sentence never gets said again. Not because the researcher got faster but because the operating model got smarter. The right questions get the right treatment at the right speed. The fast work is fast and credible. The deep work is protected and funded. The evidence is bounded, traceable, and honest about what it can and cannot support. And the researcher is no longer the person the team routes around. They are the person the team routes through.

The tools changed. The tempo changed. But the craft did not collapse. It adapted. And the researchers who built the systems described in this book are not the ones wondering whether their function is relevant. They are the ones whose calendars are full, whose stakeholders trust them, and whose organizations make better decisions because they exist.

That is the future of the craft. It is already here for the people who built for it. This book gave you the blueprints; now build.

Acknowledgments

This book started as a blog. For a long stretch, *The Voice of User* (thevoiceofuser.com) was a place to think out loud about where research was heading, and the readers who showed up week after week, more than a thousand of you now, are the reason those half-formed ideas became something worth writing down. The Frame, the Delta, the routing logic, all of it was argued into shape in public, with you. Thank you for staying in the conversation.

To Jing Jing, thank you for backing this and for making room on the team for the kind of thinking that does not always have an immediate deliverable attached. To Lauren and Joann, thank you for clearing the path internally and for backing this work before there was a finished book to point to. Working alongside this group of cross-functional partners shaped more of these ideas than any of them realize.

To the alpha readers who gave their time without being asked twice, thank you. You caught the places where I was being clever instead of clear.

Audrey, my copyeditor, saved me from myself more times than I want to count. Sebastián gave the book a cover that actually looks like the thing it is, then quietly did everything else too, the typesetting, the layout, all the work of turning a manuscript into an actual book you can hold. Thank you both.

Some of the frameworks here have much older roots, going back to the years I spent building a research function from scratch and learning, often the hard way, what holds up under real pressure and what does not. I am grateful to everyone who let me make those mistakes on their watch.

And to my wife and children, who are the whole reason any of this matters. You put up with the early mornings and the late nights, the weekends that disappeared into chapters, and the version of me that was half-present while the other half was somewhere inside this book. You never once made me feel like it was too much to ask, even when it was. Everything good in my life traces back to you. Thank you for being home.

About the Author

Constantine Papas has spent over thirteen years helping teams build better products by understanding the people who use them. He has built research functions from the ground up, led and scaled research teams, and run studies in rooms with whiteboards, Figma prototypes, sticky notes, and, once, in a supply closet. He has seen how research gets done and undone in the wild, and he spent most of those years figuring out how to make it matter.

He has a bachelor's degree in computer science and master's and Ph.D. degrees in human–computer interaction. He has been a researcher, a research lead, a professor, and, briefly, a guy refreshing his email waiting to hear back about jobs. That last one was the most formative experience of his career.

He runs *The Voice of User* (thevoiceofuser.com), a blog he started because somewhere between endorsing a stranger for "AI Empathy Strategy" on LinkedIn and replying "Thanks for sharing!" to a VP he once saw fire a whole team via spreadsheet, he knew he was too far gone. It is where he writes opinionated, longform takes on UX research, product decision-making, and the ongoing tension between what we know and what we choose to do anyway. If you enjoyed this book's tone and want more of it in shorter doses, that is where to find it.

He lives in New York with his wife, who has shown extraordinary patience with a man who writes about research governance at midnight; his two kids; and two dogs that have shown no patience with anything, ever.

You can reach him at cpapas@thevoiceofuser.com.

www.ingramcontent.com/pod-product-compliance
Lightning Source LLC
LaVergne TN
LVHW022201200926
848481LV00006B/1805
* 9 7 9 8 9 9 6 7 8 5 6 0 5 *